APPLICATION OF PROBABILISTIC METHODS FOR THE SAFETY ASSESSMENT AND THE RELIABLE OPERATION OF RESEARCH REACTORS

The following States are Members of the International Atomic Energy Agency:

AFGHANISTAN
ALBANIA
ALGERIA
ANGOLA
ANTIGUA AND BARBUDA
ARGENTINA
ARMENIA
AUSTRALIA
AUSTRIA
AZERBAIJAN
BAHAMAS
BAHRAIN
BANGLADESH
BARBADOS
BELARUS
BELGIUM
BELIZE
BENIN
BOLIVIA, PLURINATIONAL
   STATE OF
BOSNIA AND HERZEGOVINA
BOTSWANA
BRAZIL
BRUNEI DARUSSALAM
BULGARIA
BURKINA FASO
BURUNDI
CAMBODIA
CAMEROON
CANADA
CENTRAL AFRICAN
   REPUBLIC
CHAD
CHILE
CHINA
COLOMBIA
COMOROS
CONGO
COSTA RICA
CÔTE D'IVOIRE
CROATIA
CUBA
CYPRUS
CZECH REPUBLIC
DEMOCRATIC REPUBLIC
   OF THE CONGO
DENMARK
DJIBOUTI
DOMINICA
DOMINICAN REPUBLIC
ECUADOR
EGYPT
EL SALVADOR
ERITREA
ESTONIA
ESWATINI
ETHIOPIA
FIJI
FINLAND
FRANCE
GABON
GEORGIA

GERMANY
GHANA
GREECE
GRENADA
GUATEMALA
GUYANA
HAITI
HOLY SEE
HONDURAS
HUNGARY
ICELAND
INDIA
INDONESIA
IRAN, ISLAMIC REPUBLIC OF
IRAQ
IRELAND
ISRAEL
ITALY
JAMAICA
JAPAN
JORDAN
KAZAKHSTAN
KENYA
KOREA, REPUBLIC OF
KUWAIT
KYRGYZSTAN
LAO PEOPLE'S DEMOCRATIC
   REPUBLIC
LATVIA
LEBANON
LESOTHO
LIBERIA
LIBYA
LIECHTENSTEIN
LITHUANIA
LUXEMBOURG
MADAGASCAR
MALAWI
MALAYSIA
MALI
MALTA
MARSHALL ISLANDS
MAURITANIA
MAURITIUS
MEXICO
MONACO
MONGOLIA
MONTENEGRO
MOROCCO
MOZAMBIQUE
MYANMAR
NAMIBIA
NEPAL
NETHERLANDS
NEW ZEALAND
NICARAGUA
NIGER
NIGERIA
NORTH MACEDONIA
NORWAY
OMAN
PAKISTAN

PALAU
PANAMA
PAPUA NEW GUINEA
PARAGUAY
PERU
PHILIPPINES
POLAND
PORTUGAL
QATAR
REPUBLIC OF MOLDOVA
ROMANIA
RUSSIAN FEDERATION
RWANDA
SAINT KITTS AND NEVIS
SAINT LUCIA
SAINT VINCENT AND
   THE GRENADINES
SAMOA
SAN MARINO
SAUDI ARABIA
SENEGAL
SERBIA
SEYCHELLES
SIERRA LEONE
SINGAPORE
SLOVAKIA
SLOVENIA
SOUTH AFRICA
SPAIN
SRI LANKA
SUDAN
SWEDEN
SWITZERLAND
SYRIAN ARAB REPUBLIC
TAJIKISTAN
THAILAND
TOGO
TONGA
TRINIDAD AND TOBAGO
TUNISIA
TÜRKİYE
TURKMENISTAN
UGANDA
UKRAINE
UNITED ARAB EMIRATES
UNITED KINGDOM OF
   GREAT BRITAIN AND
   NORTHERN IRELAND
UNITED REPUBLIC
   OF TANZANIA
UNITED STATES OF AMERICA
URUGUAY
UZBEKISTAN
VANUATU
VENEZUELA, BOLIVARIAN
   REPUBLIC OF
VIET NAM
YEMEN
ZAMBIA
ZIMBABWE

SAFETY REPORTS SERIES No. 107

APPLICATION OF PROBABILISTIC METHODS FOR THE SAFETY ASSESSMENT AND THE RELIABLE OPERATION OF RESEARCH REACTORS

INTERNATIONAL ATOMIC ENERGY AGENCY
VIENNA, 2023

COPYRIGHT NOTICE

© IAEA, 2023

Printed by the IAEA in Austria
May 2023
STI/PUB/1946

IAEA Library Cataloguing in Publication Data

Names: International Atomic Energy Agency.
Title: Application of probabilistic methods for the safety assessment and the reliable operation of research reactors / International Atomic Energy Agency.
Description: Vienna : International Atomic Energy Agency, 2023. | Series: IAEA safety reports series, ISSN 1020–6450 ; no. 107 | Includes bibliographical references.
Identifiers: IAEAL 23-01568 | ISBN 978–92–0–111421–1 (paperback : alk. paper) | ISBN 978–92–0–111521–8 (pdf) | ISBN 978–92–0–111621–5 (epub)
Subjects: LCSH: Industrial safety. | Nuclear reactors — Safety measures. | Nuclear reactors — Risk assessment.
Classification: UDC 621.039.58 | STI/PUB/1946

FOREWORD

IAEA Safety Standards Series No. SSR-3, Safety of Research Reactors, establishes the requirements for the use of probabilistic methods to complement deterministic safety analysis. Probabilistic methods are increasingly being used in countries operating research reactors as a complementary tool to assess the risks associated with a range of potential accident scenarios. By identifying the adverse effects of various risk contributors, measures to further enhance research reactor safety can be determined and incorporated into the decision making process.

This publication provides practical information on the application of probabilistic methods for the safety assessment and the reliable operation of research reactors. It addresses features specific to research reactors and suggests an approach for the development and implementation of a project using probabilistic methods in terms of objective, scope, data and modelling, as well as the application of results to enhance safety and reliability. This publication is intended to be used by operating organizations, regulatory bodies and technical support organizations when performing or reviewing research reactor assessments in which probabilistic methods are applied. It will ideally be read in conjunction with relevant IAEA Safety Standards Series publications and technical guidelines for safety analysis, operation and maintenance, and component reliability data for research reactors. The supplementary files, available on-line, provide case studies from Member States of applying probabilistic methods for the assessment and operation of research reactors.

The IAEA greatly appreciates the contributions of all those who were involved in the drafting and review of this publication. The IAEA officers responsible for this publication were A. Shokr of the Division of Nuclear Installation Safety and R. Sharma of the Division of Nuclear Fuel Cycle and Waste Technology.

EDITORIAL NOTE

CONTENTS

1. INTRODUCTION

1.1. BACKGROUND

The deterministic method has been dominant in the design, operation and regulatory review and assessment of research reactors. However, probabilistic methods are increasingly being used as a complementary tool, allowing for a risk informed decision making process to assess the safety and reliability of research reactors. This publication addresses the application of probabilistic methods to research reactors, mainly in the areas of probabilistic safety assessment (PSA) and reliability, availability and maintainability (RAM) analysis.

Requirement 41 of IAEA Safety Standards Series No. SSR-3, Safety of Research Reactors [1], states:

"**A safety analysis of the design for a research reactor facility shall be conducted in which methods of deterministic analysis and complementary probabilistic analysis as appropriate shall be applied to enable the challenges to safety in all facility states to be evaluated and assessed.**"

Paragraph 6.123 of SSR-3 [1] states:

"For each accident sequence considered, the extent to which the safety systems and any operable process systems are required to function under accident conditions shall be indicated. These events are usually evaluated by deterministic methods. Probabilistic techniques can be used to complement the evaluation. The results of these complementary analyses shall provide input to the design of the safety systems and the definition of their functions."

Requirement 15 of IAEA Safety Standards Series No. GSR Part 4 (Rev. 1), Safety Assessment for Facilities and Activities [2], states that "**Both deterministic and probabilistic approaches shall be included in the safety analysis.**"

Paragraph 4.55 of GSR Part 4 (Rev. 1) [2] states:

"The objectives of a probabilistic safety analysis are to determine all significant contributing factors to the radiation risks arising from a facility or activity, and to evaluate the extent to which the overall design is well balanced and meets probabilistic safety criteria where these have been defined. In the area of reactor safety, probabilistic safety analysis uses a comprehensive, structured approach to identify failure scenarios. It constitutes a conceptual and mathematical tool for deriving numerical estimates of risk. The probabilistic approach uses realistic assumptions whenever possible and provides a framework for addressing many of the uncertainties explicitly. Probabilistic approaches may provide insights into system performance, reliability, interactions and weaknesses in the design, the application of defence in depth, and risks, that it may not be possible to derive from a deterministic analysis."

In addition to PSA, RAM analyses provide input to improve the design, availability and maintenance of research reactors [3]. Paragraph 1.8 of SSR-3 [1] states:

"Research reactors with power levels in excess of several tens of megawatts, fast reactors and reactors using experimental devices such as high pressure and temperature loops and cold or hot neutron sources may require … the application of requirements for power reactors".

For PSA, it is suggested to set numerical target probabilistic goals in terms of core damage frequency (CDF) and large early release frequency [4].

The earlier IAEA publications relating to PSA for research reactors [5, 6] were published in the 1980s and 1990s and required revision in terms of enhancing their scope and making provision for the modelling requirements of the wide variety of more than 220 research reactors operating around the world. This publication updates the information in Refs [5, 6]. Other publications related to PSA for research reactors, such as IAEA-TECDOC-636, Manual on Reliability Data Collection for Research Reactor PSAs [7], and IAEA-TECDOC-1922, Reliability Data for Research Reactor Probabilistic Safety Assessment [8], can be referred to for information on reliability data for a specific research reactor facility. Reference [3] covers management system attributes and good practices to optimize the availability and reliability of research reactors.

For nuclear power plants, PSA procedures have been developed at three levels:

(1) IAEA Safety Standards Series No. SSG-3, Development and Application of Level 1 Probabilistic Safety Assessment for Nuclear Power Plants [9];
(2) IAEA Safety Standards Series No. SSG-4, Development and Application of Level 2 Probabilistic Safety Assessment for Nuclear Power Plants [10];

(3) IAEA Safety Series No. 50-P-12, Procedures for Conducting Probabilistic Safety Assessments of Nuclear Power Plants (Level 3)[1].

In the absence of specific guidance for research reactors, PSA practitioners have adopted the guidance published for nuclear power plants to perform PSA in research reactors using a graded approach. When performed at the design stage, PSA may be used to evaluate designs and optimize system configuration. During the operational phase, PSA may be used to review safety performance. Moreover, the probabilistic method may be used to conduct RAM analysis, as demonstrated in this publication.

Special modelling using the probabilistic method is needed to assess the safety and reliability of research reactors owing to the diverse characteristics of this reactor type:

(a) Varieties of design and size: critical facility, open pool type, tank type, tank in pool; multipurpose design; and power levels ranging from a few watts to hundreds of megawatts.
(b) Location — often in close vicinity to populated areas.
(c) Higher excess reactivity and power density (in some cases).
(d) Additional hazards posed by test and experimental facilities.
(e) Flexible core configurations.
(f) More frequent changes in operational modes.
(g) Fewer risk mitigation features.
(h) Frequent operator interventions.
(i) Less rigorous containment or confinement systems than nuclear power plants.

These diversities serve to encourage the application of the probabilistic method for RAM analysis, despite considerable challenges. It is recognized that the requirements of failure rate data for PSA and RAM analysis are different. For example, for an assessment of safety, only unsafe failure rates are required, while for reliability, all failure rates are required. Therefore, distinctions between PSA and RAM analysis are made when discussing specific topics.

Finally, considering the regulatory practices of Member States, it is also recognized that an increasing number of regulatory bodies now require PSA to be performed at research reactors.

---

[1] IAEA Safety Series No. 50-P-12 is obsolete. A new IAEA TECDOC is at an advanced stage of development and will provide a comprehensive discussion of Level 3 PSA.

1.2. OBJECTIVE

The primary objective of this publication is to provide practical information on the application of probabilistic methods for the safety assessment and the reliable operation of research reactors, and on the use of these methods to support their design, operation and regulation. The information provided in this publication could be used to share insights into safety through PSA and into reliability through RAM analysis.

The information provided in this publication is based on prevailing international good practice, although it does not preclude the use of equivalent or alternative approaches. Guidance provided here, describing good practices, represents expert opinion but does not constitute recommendations made on the basis of a consensus of Member States.

1.3. SCOPE

This publication covers the application of Level 1 PSA for internal hazards during full and low power (during experiments) and shutdown states, with the reactor core as the source of radioactivity. It provides general steps for Level 2 PSA and Level 3 PSA and describes the probabilistic method used for RAM analysis. The publication specifically refers to small (up to 2 MW) and medium power (up to 20 MW) research reactors. For larger research reactors, practices commonly used for nuclear power plants may be more applicable, with special attention being paid to specific research reactor features such as experimental facilities and high temperature and pressure loops. This publication also covers the development of a project using these probabilistic methods by presenting information and best practices related to defining the objective, scope, data, modelling and finally utilization of insights available. Training for the performance and use of these probabilistic methods in research reactors is also discussed. In order to highlight the application of these probabilistic methods, illustrations and practical examples are provided. This publication does not supersede any IAEA publication on the topic of probabilistic methods.

1.4. STRUCTURE

Following the introduction in Section 1, Section 2 provides information on the organization and management of probabilistic methods and includes quality assurance and the application of quality attributes. Section 3 discusses the general aspects and characteristics of Level 1 PSAs, as well as the low power and shutdown PSA method applicable to research reactors. It also discusses general aspects of Level 2 and Level 3 PSAs for research reactors. Section 4 addresses the application of probabilistic methods to support design, operation, maintenance, utilization and modifications, and discusses procedures and best practices for the application of RAM analysis for research reactors. Section 5 provides information on the training of operators, regulators and practitioners of PSA and RAM analysis on the use of probabilistic methods.

Appendix I presents a typical table of contents for a Level 1 PSAs report. Appendix II lists examples of facility operational states (FOSs) for high power research reactors. The annexes, which are available on-line only as supplementary files[2], provide case studies from Member States of applying probabilistic methods for the assessment and operation of research reactors.

2. GENERAL CONSIDERATIONS REGARDING PROBABILISTIC METHODS FOR RESEARCH REACTORS

2.1. INTRODUCTION

Safety assessments are performed to demonstrate the safety of a facility and provide adequate assurance that it has been designed and built and is being operated in line with regulatory requirements and that the probability of accidents involving a potential release of radioactivity is acceptably low. This assurance is effected through a safety analysis that takes into account all postulated initiating events (PIEs) and operating conditions. Safety analyses are also used in the development of operational limits and conditions, operating procedures, maintenance, inspection and periodic testing programmes, as well as surveillance programmes and emergency planning. Safety assessments cover the response of safety systems, engineered safety features and recovery provisions for given initiating events. These assessments are typically carried out using deterministic methods for research reactors. Probabilistic methods enable the quantification of statements of safety, reliability, availability and maintainability for systems, structures and components (SSCs). Probabilistic assessment facilitates the integration of SSC performance by using generic or plant specific data to predict the safety, reliability, availability and maintainability of respective SSCs.

---

[2] Available on the publication's individual web page at www.iaea.org/publications

Probabilistic methods — PSA for safety and RAM analysis for reliable operation of the facility — can be used to complement deterministic methods.

There are many common procedural aspects of organizing and managing a project using probabilistic methods (e.g. PSA or RAM analysis), including facility familiarization and functional analysis; selection and grouping of initiating events; systems analysis; data collection and parameter estimation; setting of probabilistic goals and criteria and comparison of facility performance with these goals; uncertainty evaluation; importance and sensitivity analysis; and document preparation. In this context, the guidelines and best practices for the organizational and management aspects of PSA described below can be suitably adapted for RAM analysis, which also adopts a probabilistic method. Nevertheless, specific aspects of PSA may differ from RAM analysis, such as the requirement for reliability data, modes of failure, and the interpretation of applicable models for availability and safety.

2.2. OBJECTIVE AND SCOPE OF A PROJECT USING PROBABILISTIC METHODS

The objective and the scope of a project using probabilistic methods need to be clearly defined. The motivation for the assessment helps to determine the aims and objectives of the project. For example, managerial stipulations could require that results of assessments using probabilistic methods be submitted to support a long term programme.

At the design stage, the project using probabilistic methods is carried out using only generic data. For an operational facility, the objective of the project is to identify and prioritize vulnerable areas for operation, modification and safety improvement.

The probabilistic methods can be used in two ways:

(1) As a tool for the relative comparison and prioritization of safety and reliability, availability and maintainability issues, without any numerical goals or criteria being set;
(2) For PSA, to obtain an absolute value of the undesired end state, such as CDF and large early release frequency.

2.3. MANAGEMENT OF A PROJECT USING PROBABILISTIC METHODS

2.3.1. Project proposal

The project using probabilistic methods can be initiated either by the operating organization or by the regulatory body. The project proposal includes the following items:

(a) Objectives and scope;
(b) Management scheme, including methods, procedures and necessary resources;
(c) Team composition;
(d) Project schedule;
(e) Quality assurance procedures;
(f) Peer review mechanism.

It is preferable to discuss the project proposal with the stakeholders, and for PSA to involve the regulatory body (if required). Human and financial resources necessary for the project, such as staff, computational facilities, software, reviews, secretarial support and any consultancy or other collaborative work, need to be considered and assessed. This evaluation of resources may take into account any analysis that has already been performed, for example, a system reliability analysis or data collection.

2.3.2. Structure of a project team

The project team is led by a project manager who is responsible for the execution of the project and has the authority to implement requirements and communicate with various stakeholders. The team consists of probabilistic method specialists, designers, operation and maintenance specialists, a human reliability expert, a data analyst and an accident analysis expert. If a team is performing a probabilistic assessment for the first time, team members need to receive the necessary training to complete the project successfully.

The project can also be outsourced to an external expert organization. In such cases, it is important that the expertise and capability of the contractor, as well as the lines of communication between the contractor and the decision making organization, are well defined. Even if the project is being carried out by an external contractor, the primary responsibility for the project remains with the organization awarding the contract.

2.3.3. Selection of working methods and procedures

Appropriate working methods and procedures have to be established at the beginning of the project. Details regarding this are elaborated in the relevant sections of this publication.

The definition of the 'reference condition' of the facility has to be established in terms of the status of systems, operating policy and regulatory stipulations at a defined date to have consistency in the input and the data coming from it. The period of data covered by the project needs to be relevant to that reference condition. For example, if a system such as a relay based trip system was replaced by a digital system ten years ago, only those ten years of data are relevant. As another example, if the performance of the emergency diesel generators has degraded in the last five years, then data covering a period much longer than five years could mask recent data and may not allow for a realistic prediction of the failure frequency of the generator.

2.3.4. Time schedules

The time schedule of the project includes major analysis elements and their start and completion times. This schedule has to take into account the dependence, overlap or interdependence of activities; collection of information on reactor systems, experimental and engineering test facilities and aspects related to fuel handling and storage; required modelling; available resources (e.g. computational infrastructure, team members' time, training needs, tools such as software); and peer review. The following information needs to be collected:

(a) Design basis reports or manuals;
(b) Operation and maintenance manuals;
(c) Operational history and equipment fault reports;
(d) Test reports and maintenance information;
(e) Human error data and insights into common causes of system failure.

2.3.5. Documentation

A systematic approach is needed to document how the input data have been obtained to ensure the traceability of the project results. The computer codes used in the analysis have to be benchmarked, validated and verified. The process of benchmarking and the acceptance criteria used have to be documented. Major assumptions in the project need to be recorded and a sensitivity study has to be carried out to investigate the effect of these assumptions.

2.3.6. Review

A review of the project results is necessary prior to submitting a report to the stakeholders. For example, PSA results are subject to internal and peer review before they are submitted for regulatory review. IAEA guidelines [11, 12] for the regulatory review of Level 1 and Level 2 PSAs, mainly targeted at nuclear power plants, can also be adopted for research reactors.

2.4. BENEFITS AND LIMITATIONS OF PROBABILISTIC METHODS

Probabilistic methods can have several benefits and limitations that have to be considered prior to embarking on the project. It is important to emphasize that a complete project using probabilistic methods for any reactor requires a substantial investment of skilled effort. While the potential benefits may readily be seen to justify the costs in the case of a nuclear power plant, particularly when several similar units may benefit from the same project, this may not always be the case for a single, unique type of research reactor with special application and utilization. In the case of the research reactor, a graded approach to the projects of limited scope and directed at specific issues may be more effective and practical to meet the objectives.

2.4.1. Benefits

Probabilistic methods have the following benefits:

(a) They are an integrated technique that incorporates many aspects of a facility into the methodology in a systematic way, namely design features, operating practices, operating history, maintenance procedures, components of RAM and human factors;
(b) They provide an explicit framework for uncertainty quantifications;
(c) They can be used for comparative purposes.

While PSA applications have grown significantly, the potential benefits that can be realized by using RAM analysis in terms of improving research reactor availability and reliability are also attracting increased attention. One of the well known methods for improving facility availability is reliability centred maintenance. This method has been widely used in many other industries and can be effective in improving research reactor availability. A reliability based method can also be applied to optimize the inspection intervals for SSCs. Moreover, considering that the research reactor fleet is ageing, a probabilistic method can be a valuable tool that supports the renewal of operational licences and ageing management programmes, as well as decision making.

2.4.2. Limitations

The limitations of probabilistic methods arise from the fact that the results carry uncertainties from different sources. The uncertainties that arise in probabilistic methods are characteristically different from those that arise in deterministic results. Where data are uncertain, relative judgements can be made for managerial purposes. If available, the use of high quality data throughout will add further confidence to the estimation, and where there is acknowledged uncertainty, the application of sensitivity analysis can identify where to concentrate on accuracy.

There are four main sources of uncertainty in any probabilistic method:

(1) Completeness uncertainties. These arise because of a lack of full data about the facility. To minimize this, the data used need to be derived from a comprehensive, integrated model of the facility. These data cover all credible scenarios to address all modes of operation of the facility.

(2) Parameter uncertainties. Such uncertainties are explicitly represented in the data source and they are available in the form of probability distributions. For PSA, these uncertainties concern component failure rates, accident or initiating event frequencies, and probabilities for common causes of failure and human error. Statistical uncertainties from the low frequency of rare initiating events, as well as from data related to human factors, are particularly large.

(3) Uncertainties with respect to data accuracy. These contribute to the inaccuracy of reliability data. They influence both generic and facility specific data and can be grouped into two major areas: differences in data collection or processing and differences in actual component reliability. In each of these areas, individual factors influencing reliability parameters can be identified.

(4) Model uncertainties. These are associated with the modelling assumptions. These uncertainties cannot be easily quantified, and include, for example, the performance of certain components in rare conditions, physical phenomena or human actions that cannot be completely predicted.

2.5. ESTABLISHING QUALITY ATTRIBUTES FOR PROBABILISTIC METHODS

To achieve standardization and harmonization of all aspects of a probabilistic method for research reactors, it is necessary that the numerical process, data collection, treatment of uncertainty, and compilation and presentation of results, including documentation, follow standard quality criteria [13–15].

These quality attributes define the best practices for the application of a probabilistic method and could be suitable for high power research reactor facilities (>20 MW). For lower power research reactors, a graded approach could be applied (with suitable justification) to adapt these attributes for the application of probabilistic methods on an adequate scale.

The framework proposed in this publication reduces variability in the probabilistic method procedure, the use of data and models, the definition of success criteria and the interpretation of the project results. As this framework brings clarity and understanding with respect to the quality of probabilistic methods, it may also help in the regulatory review of the project.

Reference [15] provides a list of the quality attributes of the full scope Level 1 PSA for nuclear power plants, covering both general attributes (applicable for the base case PSA), and application specific ones (i.e. special attributes). The quality attributes elaborated for nuclear power plants might be adopted for research reactors, taking into account their specific design features and applying a graded approach. The attributes that are required for a particular application depend on the purpose and characteristics of the application. When used as an input to a decision, the attributes required are a function of the decision making process, and in particular address the acceptance criteria or guidelines against which the PSA results are to be compared.

2.6. APPLICATION OF PROBABILISTIC METHODS FOR THE SAFETY ASSESSMENT AND THE RAM ANALYSIS

Probabilistic methods can be used in projects throughout the lifetime of a research reactor facility, with a different focus depending on the activity concerned (e.g. design, operation, maintenance, utilization or modification) and the purpose of the project (e.g. PSA for safety or RAM analysis for reliable operation). Probabilistic methods can provide useful insights and inputs for various interested parties, such as facility staff (management, engineering, operations and maintenance personnel), regulatory bodies, designers and vendors, to assist with decision making in the following areas:

(a) Research programmes;
(b) Design and facility modifications;
(c) Safety classification of SSCs;
(d) Operational limits and conditions;
(e) Routine facility operation;
(f) Review and validation of operating procedures;
(g) Maintenance, periodic testing and inspection programmes;
(h) Ageing management programmes;
(i) Accident management;
(j) Regulatory aspects and licence compliance;
(k) Evaluation of research reactor design and operation for RAM analysis;
(l) Reliability centred maintenance and inventory management;
(m) Risk management (both safety and investment).

3. PROBABILISTIC SAFETY ASSESSMENT

3.1. INTRODUCTION

Probabilistic safety assessment is a systematic and effective method to predict the risk from postulated accident scenarios for research reactors. Using these insights, risk informed applications can be developed, such as risk monitors for operational decisions, risk informed in-service inspection and maintenance management through the identification and prioritization of SSCs in need of maintenance or repair.

3.1.1. Objectives

In general terms, a PSA aims to assess the level of safety of the facility. This is achieved together with the following three objectives:

(1) To identify the most effective areas for design improvement;
(2) To compare the level of safety with safety goals;
(3) To assist facility operation and maintenance.

In practical applications, a PSA may also identify and delineate the combinations of events that may lead to undesired end states, such as severe accident and radioactive releases; assess the expected probability of occurrence for each top event in event trees; and evaluate the consequences.

3.1.2. Methodology

The PSA methodology provides a systematic and integrated framework for the quantification of risk at a research reactor facility. The basic PSA approach lies in considering that any component of the facility can 'fail'. Risk evaluation is then performed by considering the failure probabilities for the primary components.

The PSA methodology integrates information about a facility's design, operational procedures, operating history, SSC reliability or availability, maintenance procedures, human performance, accident phenomena and potential environmental and health effects. The approach aims to achieve completeness in identifying deficiencies and facility vulnerabilities, and to provide a balanced picture of the safety significance of issues, including uncertainties in numerical results. The PSA needs to take into account the potential for release from all significant sources of radioactivity at the facility, such as core fuel, spent fuel, stored radioactive waste, irradiated materials and experimental facilities.

3.1.3. Levels

Probabilistic safety assessment is an analytical approach used to postulate accident scenarios and evaluate risk in them. The generally recognized three levels of PSA [16] may apply to research reactors:

(1) Level 1 PSA deals with the system modelling aspects of the research reactor facility. Design and operational aspects are analysed to postulate accident scenarios. The output of Level 1 PSA is a spectrum of accident sequences and their associated frequencies. Using this information, the CDF of the facility is estimated. Level 1 PSA provides insights into the strengths and weaknesses of the facility's design and its operation.

(2) Level 2 PSA builds on input on the chronological progression of accident sequences from Level 1 PSA and models confinement or containment scenarios for identified facility damage states to evaluate release frequency. The modelling involves the identification of scenarios involving releases of radioactive material from the source (for research reactors, the source could be the core; stored spent fuel; and experimental, test or irradiation facilities) and the potential for releases from confinement and containment. The major

output of Level 2 PSA is radioactive release frequencies for the range of source terms.

(3) Level 3 PSA deals with the assessment of risk to members of the public, taking into account the release frequencies obtained from Level 2 PSA. This involves consideration of the conditions for the whole spectrum of radioactive release to the environment, including atmospheric dispersion and soil and water contamination. The output of Level 3 PSA is a risk statement for members of the public, property and other societal elements. Level 3 PSA also provides insights into the relative effectiveness of aspects of accident management relating to emergency preparedness and response.

3.2. SCOPE OF PROBABILISTIC SAFETY ASSESSMENT

The scope of a PSA is to be set according to the clear objectives of the analysis, as discussed in Section 2.2. Therefore, determination of the objectives of the PSA, together with the intended and potential use of its results, is an important step prior to performing PSA.

The quantitative results of PSA are often used to verify compliance with safety goals or criteria, which are set in terms of quantitative estimates of CDF (for Level 1), frequencies of radioactive releases of different types (for Level 2) and probability of societal risks (for Level 3).

Safety goals or criteria do not indicate which hazards and operational modes have to be addressed. Therefore, in order to use the PSA results for the verification of compliance with safety goals or criteria, full scope PSA involving a comprehensive list of initiating events and hazards for all operational modes needs to be performed. The common practice is to perform the analysis for the various internal and external hazards and operational modes in separate modules, with Level 1 PSA for full power operating conditions using internal initiating events as a basis, as described in SSG-3 [9].

To determine the objectives of the analysis and thereby the scope of a PSA, the aspects discussed below have to be considered.

3.2.1. Analysis level

The first consideration is PSA level. If the analysis is only carried out to Level 1, the focus of the analysis is usually the reactor core, in-pile experiments and fuel storage. If the PSA is carried out to Level 2 or Level 3, where the impact of radioactive release is assessed, the scope of the PSA includes additional contributions to risk arising from other radioactive material at the site, such as irradiated materials and irradiated experimental facilities, as well as radioactive waste.

3.2.2. Stages of facility lifetime

The second consideration is the stage of the facility lifetime. A PSA can be performed to analyse the different stages of the facility lifetime:

(a) Siting;
(b) Design;
(c) Construction;
(d) Commissioning;
(e) Operation (including modification or refurbishment);
(f) Decommissioning.

A PSA can have both general and specific objectives for any of the given lifetime stage(s). Nevertheless, it is desirable to start the PSA process as early as possible in the lifetime of the facility, allowing any identified design weaknesses to be corrected or improved in a cost effective manner.

In the evaluation of the suitability of a site for a research reactor during the siting stage, site characteristics that may affect safety aspects of the reactor have to be investigated and assessed. These include environmental characteristics of the region that may be affected by potential radioactive releases from the reactor in operational state and in accident conditions.

Assessment techniques such as failure mode and effect analysis, fault trees, event trees and reliability block diagrams can be used to assist system level analysis prior to the development of a complete PSA, to provide technical support for design development, to compare alternative designs or to verify compliance with probabilistic targets (reliability or availability).

3.2.3. Potential sources of radioactive release

The third consideration is the requirement to review all potential radioactive sources or radiation fields in a research reactor. These include the following:

(a) Fission product inventory of the reactor core fuel;
(b) Fuel in test loops;
(c) Spent fuel in storage;
(d) Fission products and activation products in the pool or coolant system and in related systems, such as purification systems;
(e) Equipment, systems and piping containing radioactive material;

(f) Solid and liquid waste and waste management facilities, and leakage or spills from these facilities;

(g) Gaseous radioactive materials from the pool, coolant systems, cover gas systems, reflector systems and experimental facilities connected to ventilation systems or any leakage from these systems;

(h) Filters of the ventilation systems;

(i) Airborne radioactive materials in areas normally occupied by personnel;

(j) Experimental and production facilities with the potential to generate activated or other radioactive material, or facilities for the storage and handling of such material, including sample irradiation facilities, in-core experiments and hot cells;

(k) Materials irradiated by the reactor;

(l) Neutron startup sources;

(m) Sources for the testing and calibration of radiation monitoring equipment.

Usually, the radionuclide inventory in experimental devices is much lower than the radionuclide inventory of the reactor core. Therefore, the hazard associated with failures of experimental devices will be considerably lower than the hazard associated with fuel failures in the reactor core. However, experimental devices and features other than the core facility SSCs (e.g. graphite reflectors) may pose a risk to experimenters and operators of the reactor, for reasons such as the following:

(a) Experiments often change and vary in their set-up.

(b) Irradiation facilities may involve rapidly changing levels of activity in the inventories of radioactive materials, or the levels of activity may be unknown or may not have been estimated.

(c) Because of their temporary nature, experiments and irradiation facilities usually have a higher probability of failure than the reactor systems and components themselves.

(d) Experimental devices are often located in proximity to the operators and inadequate physical barriers may have been provided for the protection of experimenters and operating personnel.

(e) For some out-of-core experimental facilities, no physical barriers are installed and protection for personnel is solely dependent on compliance with radiation protection procedures.

Sources of radioactive material release from experimental facilities of research reactors can include the following:

(a) Irradiation of material (e.g. ^{99}Mo production from fissile targets, fuel rod testing under transient conditions up to or beyond failure);
(b) Frequent handling operations (e.g. loading and unloading of targets, sampling at glove boxes);
(c) Transfer of radioactive material from the reactor core to an out-of-core experimental device (e.g. on-line analysis of fission gas release during post-irradiation examination);
(d) Excessive ^{41}Ar production;
(e) Direct radiation from neutron beam ports;
(f) Production and transportation of radionuclides (e.g. H_2 or D_2 cold neutron sources or in-core ^{3}He circuits with out-of-pile control that may transport an appreciable amount of tritium).

For any combination of these sources of radioactive material release, a PSA may be useful to evaluate the potential hazard.

3.2.4. Facility operational states

The fourth consideration refers to the operational states of the facility at the time the event is initiated. The FOSs are divided into two main groups: normal operation and anticipated operational occurrences. Normal operation may include the following operational states:

(a) Startup;
(b) Nominal full power;
(c) Pulsed power;
(d) Low power operation;
(e) Shutdown (including extended shutdown).

3.2.5. Accident initiating events

The fifth and last consideration is the type of initiating event. In addition to internal and external hazards [1], specific aspects of research reactors, such as experimental facilities, absence of containment in many reactors, open cores in the tank or pools and storage of spent fuel close to the core, may present additional potential accident initiators.

Internal hazards originate from sources located at the facility site. Examples of internal hazards are internal fires and floods caused by internal facility failure (e.g. electric overloads or pipe ruptures), loss of off-site power, malfunctions of facility SSCs, transportation accidents including experimental facilities, release of toxic substances stored on-site and human error.

External hazards are those related to natural phenomena or human activities that do not stem from facility operation. Experience indicates that external hazards are not to be excluded as a group, but need to be included to provide a complete picture of facility risk. A preliminary screening of the external events to be considered in a PSA of a research reactor can be performed using a detailed hazard categorization. For example, for facilities in a low hazard category, some extreme scenarios (e.g. aircraft crash, tornadoes) could be screened out either by site characteristics or because of a low probability of occurrence. The screening process can then consider the potential off-site and on-site consequences induced by external events that can have an impact on research reactors [17].

After the accident at the Fukushima Daiichi nuclear power plant in Japan, which was caused by an earthquake and subsequent tsunami, the importance of performing safety assessments for all credible combinations of external hazards specific to the facility site became evident [18].

Deliberate human-caused threats, including sabotage, terrorist action or armed attacks, are also external events, but they are generally not included in a PSA unless special consideration of such acts is required.

3.2.6. Information alignment

Once all of the above aspects have been considered, information resources regarding necessary procedures, methods, personnel, expertise, funding and analysis time need to be updated and aligned to reflect the extent to which each aspect is included in the PSA.

3.3. MAJOR TASKS IN LEVEL 1 PROBABILISTIC SAFETY ASSESSMENT

3.3.1. Background

This section describes a procedure to develop a full power Level 1 PSA for a research reactor. This procedure differs from Level 1 PSA for a nuclear power plant, where the project is defined as an assessment of failures leading to the determination of CDF. The overall approach and methodology needs to be capable of modelling the fault sequences that could occur, starting from an initiating event, and of identifying the combinations of safety system failures, support system failures and human errors that could lead to core damage. For a research reactor, the analysis needs to be adapted suitably to include potential radioactive releases from other sources in addition to the reactor core, and to take into account the specific aspects of the facility design.

3.3.2. Identification of potential sources of radioactive release

In order to identify the potential sources of radioactive release[3], a list of all sources of radioactivity in the reactor and associated facilities needs to be prepared. The list has to include the type (content and form) and the activity of the radioactive materials. For example, for a pool type research reactor, the list of sources may include the reactor core content, spent fuel, handling equipment and facilities, waste storage tanks or ancillary pools, irradiation targets for radioisotope production, hot cells, and experimental facilities. If an existing source is not included in the PSA, a detailed rationale has to be provided.

3.3.3. Definition of damage states

Damage states that are defined in sufficient detail during Level 1 PSA can be used as input for Level 2 PSA. The result of Level 1 PSA (i.e. end state frequencies) depends on the definitions of damage states provided at this point. The definitions have to include all undesirable consequences. The set of end states can be presented as follows:

(a) Success. The core is intact and the reactor in a safe shutdown state is subcritical with assured means for long term cooling.
(b) Core damage. Paragraph 5.42 of SSG-3 [9] states that "A criterion (or criteria, if appropriate) should be developed for what constitutes core damage or a particular degree of core damage". For example, a criterion can be that core damage occurs if critical heat flux or flow instability occurs on fuel cladding. Specific end states are design dependent and have to be determined through analysis in each case. Determination of the extent of core damage in terms of percentage of fuel failure will provide input for Level 2 PSA.
(c) Consequence. Sometimes an additional end state is defined that is a subset of successful shutdown that has no core safety implications but could have a detrimental economic impact on facility operation or affect the exposure of operators, such as a prompt critical transient with no significant energy

---

[3] For examples of possible radiation sources or radiation fields in a research reactor, see Annex IV of IAEA Safety Standard Series No. SSG-20 (Rev. 1), Safety Assessment for Research Reactors and Preparation of the Safety Analysis Report [19].

deposition in the fuel. For non-core radioactive sources, an end state involving breach of a safety barrier may be defined.

For low power (or small radioactive inventory) facilities, a possible set of end states could be one that distinguishes whether or not the fuel or the core is damaged and whether or not it is covered by water. A typical set of end states for a pool type research reactor can be presented as follows:

(a)   End state 1. Core is intact and covered by water, reactor is in a safe shutdown state, with assured means for long term cooling and subcriticality (success).
(b)   End state 2. Core is intact but uncovered. This state could be valid for some very low power reactors in which core cooling could be achieved by air convection (success).
(c)   End state 3. Core is damaged and covered by water. This will include scenarios such as reactivity excursion or even mechanical damage to the core inside the pool (failure).
(d)   End state 4. Core is damaged and uncovered. This would be the worst scenario from the point of view of potential dose to operators or eventual releases (failure).

Each facility may need to define its unique end states prior to the analysis. Moreover, the end states are to be evaluated for their potential to contribute to large early release frequency, especially in the case of high power research reactors.

3.3.4. Selection, screening and grouping of initiating events

A number of steps are needed to complete the selection, screening and grouping of initiating events [1]. The result of this multiple step task is a list of groups of initiating events that have to be quantified and incorporated into the PSA. To achieve this goal, a list of (ungrouped) initiating events, as complete as possible, has to be generated. Since consequences are not limited to core damage, initiating events have the potential to lead to undesired end states. The list of initiating events in a PSA is not confined to the list of events postulated in the design basis. The aim is to generate a list that is as comprehensive as possible.

Initiating events that form the initial list can be selected using one or more approaches:

(a)   Evaluation. Engineering evaluation of the SSCs to identify whether any of their failure modes could lead to the undesired end states directly or in combination with other failures.

(b)   Reference lists. Lists of generic initiating events for research reactors can be found in SSG-20 (Rev. 1) [19] and obtained indirectly through the IAEA Incident Reporting System for Research Reactors [20].

(c)   Deductive analyses. These include top down analyses (similar to fault trees), in which the top event could be an undesired end state, which is branched down into all possible events that can cause this state to be reached without including the action of mitigating systems. An example of this technique is the master logic diagram described in detail in Refs [21, 22].

(d)   Operational experience. This includes experience in either the facility being analysed or in similar facilities, including analysis of incidents and near miss events.

(e)   Source and event analysis. This technique makes use of the results from the task described in Section 3.3.2. The source and event analysis method starts with the identification of radiation sources and the barriers that separate these sources from the public or occupational workers. It then postulates all credible failure mechanisms for these barriers to identify the initiating events that could cause these failures.

The next step in this task is the screening of initiating events. The initial list of initiating events has to be reviewed to remove repetition or overlap. The criteria for screening out events can be qualitative, quantitative, or both. Qualitative justification is valid, as it uses engineering judgement and takes into account the special characteristics of a particular facility. However, it requires at least a preliminary assessment of the frequency of occurrence of an initiating event and the definition of a cut-off criterion for very low frequency. Nevertheless, low frequency initiating events that could challenge mitigating systems, which have a high probability of failure, are not to be discarded.

Grouping is the process that completes the analysis of the initiating events. Initiating events that impose essentially the same success criteria on safety and mitigating systems, including the same challenges to the operators (if applicable), are grouped together. The key reason for grouping is to simplify the analysis by using the same event trees and fault trees for the initiating events in the group. Some categories of initiating events may be subdivided rather than grouped because of the distinction of required responses from SSCs important to safety. A typical example of such an event is a loss of coolant accident (LOCA), which is frequently divided by size (small, medium, large) or by location.

All initiating events have to be examined for their potential for combined or consequential occurrences, such as earthquake and induced tsunami, earthquake and induced fire, fire induced by short circuits and flooding induced by process system failure. Extreme hazards, such as an earthquake causing a tsunami, which in turn causes power loss and flooding, are unfortunate but real examples that have occurred in the past [18]. All potential occurrences of such events need to be addressed during the PSA:

(a) During analysis of initiating events, to consider the root cause as an initiator;
(b) During event tree and fault tree analysis, to consider the possibility of losing several barriers in the protection of core integrity.

3.3.5. Accident sequence modelling

A key step in building a PSA model is accident sequence modelling. In this task, a model is created that defines (i) the initiators of potential accidents, (ii) the response of the dedicated safety systems and expected human recovery actions for these initiators, and (iii) the spectrum of resulting end states.

3.3.5.1. Event tree analysis

This part of the analysis starts with each group of initiating events and determines the response of the safety systems, systematically describing whether they succeed or fail to provide the expected action.

Event tree analysis is the method widely used to model these sequences. Event trees are graphical inductive models that reflect the evolving sequence of each group of initiating events. Event tree headings reflect the status of the system or operator action, and are normally arranged in chronological order to reflect the order in which they are expected to intervene in an accident sequence. The tree displays the functional dependencies between the headings (e.g. where the failure of one system implies that another system cannot perform its function successfully). Such dependencies result in omitted branch points. Omitted branch points also occur if the failure of a given system does not affect the end state associated with a given accident sequence.

System failures are represented by another set of logical models known as fault tree analysis, which is discussed in Section 3.3.7 of this publication.

The combined event tree analysis and fault tree analysis methods are developed and presented in a variety of approaches, with the following two being the most commonly applied:

(1) The small event tree and large fault tree approach, in which dependencies between safety systems and their support systems do not appear in the event trees;
(2) The large event tree and small fault tree approach, in which dependencies between safety systems and their support systems appear in the event trees.

The small event tree and large fault tree approach generates concise event trees that allow for a synthesized view of the accident sequence. However, dependencies on the support systems are not explicitly apparent. By contrast, the large event tree and small fault tree approach explicitly shows the support systems but generates a large number of small fault trees for each safety and mitigating system, with different boundary conditions. Alternatively, the number of fault trees can be reduced by creating larger fault trees that are conditional on the state of the support systems. Both approaches are acceptable.

3.3.5.2.   Human reliability analysis

References [9, 23–26] provide information on how to model human actions and integrate human reliability analysis into PSA. A brief discussion on human reliability analysis is provided below.

Human interaction can affect both the cause and the frequency of an event sequence before, during or after the initiation of the event, and can mitigate or worsen an accident. Accordingly, this can be treated in individual components of the accident sequence modelling. The following classification scheme is therefore suggested:

(1)   Type 1. Before an initiating event, facility personnel can affect availability and safety by inadvertently disabling equipment during testing, maintenance or calibration. The probability of leaving components in an inoperable condition (e.g. misaligned valves) or their unavailability is added to other contributions at the level of basic component inputs in the fault trees. Particularly important are actions that result in the concurrent failure of multiple items that are important for safety and that contribute to common cause failures. Such items need to be coordinated closely with the analysis of common cause failures to avoid the double counting of multiple failures and to connect them properly to the logical structure of the fault tree.

(2)   Type 2. By committing an error (e.g. by not correctly following a testing and calibrating procedure), facility personnel can initiate an accident. Such events can usually be found in the facility outage database, but are not always identified as having a specific human cause. Since they are identified as initiating events, such errors are accounted for through their contribution to initiating event frequencies. Alternatively, it is assumed that initiating event frequencies already contain contributions from such errors. Most important are errors that not only precipitate the initiation of an accident but also concurrently cause the failure of items that are important to safety. Particular emphasis on common cause failures that are caused by human error is necessary.

(3)   Type 3. By following operating procedures during the course of an accident, facility personnel can terminate the accident (e.g. by operating standby equipment). This type of human action in following procedures or rules in response to an accident sequence is incorporated explicitly into fault and event trees by the systems analysts.

(4)   Type 4. Facility personnel, in attempting to follow or disregard procedures, can make an error that worsens the situation or fails to terminate accident progression. These are usually the errors of commission that occur during type 3 and 5 interactions, and are the most difficult to identify and model. Human reliability analysts and systems analysts can only identify these interactions by iteration. One example of such an interaction is when the operator misinterprets the actual state of the facility and takes actions appropriate to a different state. Another example is when the operator correctly diagnoses the event, but chooses a non-optimal strategy for dealing with it. Once the probable actions of this type have been identified, they can be incorporated into an event tree. Only a few PSAs have attempted to include this type of interaction, and even these only to a limited degree. Appropriate data are not available for predicting these types of human interaction. However, a retrospective analysis of actual events can usually identify those interactions that have occurred in the facility.

(5)   Type 5. During the course of an accident, facility personnel can restore and operate initially unavailable equipment to terminate the accident. These interactions consist of recovery actions that are included in accident sequences that dominate risk profiles and may include the recovery of previously unavailable equipment or the use of non-standard procedures to improve the accident conditions. They can be incorporated into the PSA as recovery factors in the frequency of the accident sequence cutsets[4].

Type 3, 4 and 5 actions all concern operator response to an accident once it has been initiated, and the common approach is to consider the three types as one category. Consequently, the human reliability analysis will in general have to address three categories of interaction:

(a)   Category A (type 1). Interactions that concern errors made before an accident sequence started.

(b)   Category B (type 2). Interactions that cause initiating events and especially those that concurrently cause items to fail that are important to safety.

---

[4] Cutsets are combinations of initiating events and failures and/or human errors that lead to core damage [9].

(c)   Category C (types 3, 4 and 5). Interactions that concern response to an accident sequence.

3.3.5.3.   *Classification of accident sequences into various end states*

Accident sequences are to be classified into end states. Each accident sequence (i.e. a group of initiating events followed by the combination of success or failure of the applicable mitigating system and/or human action) is assigned to one, and only one, end state. This classification process is based on deterministic analysis of the facility response to the initiating event followed by safety systems failure(s) and/or human action.

The set of end states (together with their frequency of occurrence) is the output of Level 1 PSA and is eventually utilized as the interface between Level 1 PSA and Level 2 and Level 3 PSAs. The practical approach is to group together those sequences whose end states are sufficiently alike to justify their treatment as a group during Level 2 PSA (if the PSA is to cover all three PSA levels). This approach is also applicable to the projects limited to Level 1 PSA but completed with dose calculations (or other limited consequence analysis), which is the case in many PSAs for research reactors.

The following specific considerations can help analysts in defining the set of end states (some of them may or may not apply, depending on the reactor type):

(a)   Early core damage versus late core damage (relative to time of scram);
(b)   Confinement or containment failed prior to or after core damage (both structural failure and isolation or reconfiguration failure have to be considered);
(c)   Availability of confinement or containment ventilation, pressure suppression and heat removal features;
(d)   Condition of reactor (core flooded or dry).

3.3.6.   Data compilation and analysis

The uncertainty of available data specific to research reactors is one of the key issues that deserves attention when developing PSAs. Data may not be available for one of a kind systems that need to be analysed or tested. The most common approach for estimating the parameters of the PSA models is the use of generic databases. Numerical data are needed for the following parameters:

(a)   Component failure rates;
(b)   Component failure probability on demand;
(c)   Component mean time to repair;

(d) System and component test frequencies;
(e) System and component test duration;
(f) System and component maintenance frequencies;
(g) System and component maintenance duration;
(h) Common failure data and model;
(i) Human reliability data and model.

Reference [8] could be applied for the preliminary PSA, or for PSAs that are developed during the design stage of a research reactor. Once the reactor starts operation, the facility will ideally collect its own specific data. Data collected during operation can be used to update the PSA model inputs and the assessment results. More information on the collection and analysis of facility specific data is available in Ref. [7].

3.3.7. System modelling

Systems analysis can be performed using many methods, such as fault tree analysis, reliability block diagrams and Markov chain analysis. Fault trees are the most common method for modelling the failure of the safety and mitigating systems. The method comprises a deductive (top down) analysis that postulates an undesired top event and identifies all credible ways in which the top event can occur. The terminal points in the fault tree are called basic events and represent human error and the failure or unavailability of components.

The following issues need to be considered in relation to the development of fault tree models:

(a) Methods and procedures for the construction of fault trees have to be agreed at the beginning of a PSA and followed by all analysts. This is necessary in order to ensure the consistency of the analysis. Items to be considered in this context are system boundaries, logic symbols, event coding and modelling of human errors and common cause failures.
(b) All assumptions made in the process of constructing a fault tree have to be documented, together with the sources of all information used. In this way, consistency is ensured throughout the analysis and traceability is maintained.
(c) When systems are not modelled in detail and system level reliability data are used, common cause failures are to be separated out. However, this practice risks overlooking important dependencies on support systems.
(d) Clear and precise definitions of system boundaries need to be established before the analysis begins. These definitions have to be adhered to during the analysis and have to be included in the final documentation on systems

modelling. The following are examples of the definition of the interface points between safety systems and various support systems:

(i)   For electrical power supply, at the buses from which power is supplied to the components considered within the system;

(ii)   For actuation signals, at the appropriate output cabinets of the actuation system;

(iii)   For support systems providing various media (e.g. water, oil, air), at the main header line of the support system.

(e)   A standardized format has to be used for coding basic events in the fault trees. The scheme is normally selected to be compatible with the PSA software used. Common elements included in the coding of basic events are the following:

(i)   Component failure mode;

(ii)   Component type;

(iii)   Component specific label or identification;

(iv)   System to which the component belongs.

(f)   Dependent events have to be modelled explicitly and implicitly as reflected in the following points:

(i)   Multiple failure events, for which a clear cause–effect relation can be identified, have to be modelled explicitly in the system model. The root cause events have to be included in the system fault tree so that no further special dependent failure model is necessary.

(ii)   Multiple failure events that are susceptible to dependencies, and for which no clear root cause event can be identified, are commonly grouped under the name of residual common cause failures. These failures can be modelled using implicit methods such as beta factors or other parametric models [8].

(g)   For the proper quantification of accident sequences in which the initiating event may affect the operability of a mitigating system, the impact of the event on the operability of the system has to be explicitly included in each system fault tree.

(h)   To simplify and reduce the size of the fault trees, certain events can be excluded because of their low probability in comparison with other events. Examples of simplifying assumptions include the following:

(i)   Flow diversion paths for fluid systems only have to be considered if they could seriously degrade or fail the system.

(ii)   Spurious control faults for components after initial operation only have to be considered if the component is expected to receive an additional signal during the course of the accident to readjust or change its operating state.

(iii) Position faults prior to an accident are not included if the component receives an automatic signal to return to its operable state under accident conditions.

(i) The testing procedures of the components included in the system models have to be closely examined to verify whether they introduce potential failure modes. All such potential failure modes identified have to be documented. An example of such a failure would be if, in the course of testing, the flow path through a valve is isolated, and at the end of the test the flow path remains closed (possibly owing to human error) and the failure remains dormant. References [21, 22, 27] are recommended for further information on systems analysis.

3.3.8. Quantification of accident sequences

In this task, the accident sequences are determined, qualitatively analysed to verify their validity, and quantified to provide numerical results.

To determine the accident sequences, the fault trees are combined with the event trees to produce the minimal cutsets. These minimal cutsets contain the initiating event and combinations of basic events that lead to core damage. Dependencies among the initiating event, component failure and human error have to be considered during the qualitative analysis of cutsets.

Such dependencies include the following:

(a) Common cause initiating events that also cause failure in mitigating systems or support systems;
(b) Single failure modes that contribute to the failure of more than one system (shared individual faults);
(c) Dependencies caused by shared support systems;
(d) Dependencies caused by support systems embedded in another support system or in a safety system;
(e) Logic loops caused by mutual dependence of support systems on each other;
(f) Dependencies caused by the requirement to distinguish between early and late system failures.

Accident sequences also depend on whether a particular safety system fails early on during an accident or later, after the accident has been partially mitigated. For the cases involving safety system failure during an accident, it is necessary to distinguish between early and late failures of the systems. In some cases, the early failure of a system precludes any situation for which the system will be called upon later. To express this specific type of dependency on the event tree, those branches that include early failure do not branch to late failure of the same system. Support systems can also fail early on or later, resulting in accident sequence cutsets that can include both early and late failures of support systems. The early and late failures of support systems are excluded from sequences in which both early and late frontline system failure is not possible. In this case, excluded combinations of early and late failure need to be correctly accounted for.

Once the fault trees are merged with event trees, the Boolean reduction process of the event trees becomes complex, depending on the number of terms (cutsets) in the individual fault trees that make up the sequences. Thus, obtaining a numerical solution for the Boolean logic can be difficult in terms of computing requirements. The modularization of fault trees is a common approach to reduce the number of terms in the sequences.

These fault tree modules are independent subtrees that contain several primary events and are not repeated anywhere else in the model. The Boolean equation for the fault tree is then written in terms of the modules and contains considerably fewer terms than the Boolean equation written in terms of primary events, which makes the Boolean reduction process efficient. Software packages for PSA provide an option for modularization before the integrated logic is to be quantified.

3.3.9. Assessment of end state frequencies

The quantification of end state frequencies consists of the addition of all sequence frequencies that result in the same end state. The process starts by incorporating initiating event frequencies as a multiplier to each sequence on the event tree. The process is simpler for the small event tree approach, in which no special manipulations are needed.

In most cases, quantitative assessment can be accomplished in two well defined steps: initial screening and final (more refined) quantification.

For initial screening, conservative values or screening values can typically be used for basic events, including human errors, in which detailed data are not initially available. If the conservative values result in significant contributors, they need to be more precisely evaluated.

Preliminary point estimates of the frequencies of the accident sequence are calculated by multiplying the point value probability of each event tree sequence by the point value frequency estimate for the corresponding initiating event. The probability of the event tree sequence is estimated by adding together the point value failure probabilities of the component level minimal cutsets for the sequence. Post-accident recovery, such as recovery of actuation faults or of pre-accident mispositioning faults, is not credited at this stage.

Upon completion of the initial screening, the cutsets at both the system and sequence levels have to be reviewed. The purpose of this review is to verify their validity (i.e. to find any modelling errors missed in the previous reviews) and determine the significant contributors that need to be analysed further.

In order to make the sequence quantification practical, it is generally necessary to truncate the analysis (i.e. to consider only those cutsets whose probability is above a designated cut-off value, which is termed 'truncation value'). Truncation can be used in both the initial screening and the final quantification. Practice has shown that a truncation value that is lower by a factor of 1000 than a dominant value that is obtained or a criterion value that is considered is generally adequate.

The final quantification of end state frequencies completes the numerical assessment of a Level 1 PSA. This task includes the following:

(a) Requantification of the sequences that were not discarded by truncation;
(b) Inclusion of the results from detailed calculations of human errors and other basic events that were identified as significant contributors;
(c) Inclusion of recoveries, to give credit to human actions that will eliminate one of the faults in the cutsets, thus preventing the end state from occurring, by taking into account the associated probability of human error in the recoveries.

Human action can only be credited when the information necessary for the operating personnel to make the decision to act is presented clearly and unambiguously in emergency operating procedures or facility operating procedures. The operating personnel has to have sufficient time to make a decision and to act, and the physical environment has to allow the action by the operator.

3.3.10. Uncertainty, sensitivity and importance analysis

The analysis of uncertainties is conducted in two ways:

(1) Qualitative analysis that addresses the rationale and impact of uncertainties introduced in the modelling of accident sequences (e.g. uncertainty in the physical processes occurring during the accident) and in the modelling of system failures (e.g. incompleteness or simplifications made in developing the fault trees);
(2) Quantitative measures of uncertainty, which propagate from the statistical uncertainty in the input parameters.

Three major categories or sources of uncertainty in the PSA models are as follows:

(1) Completeness. The main feature of the PSA model is the assessment of the possible scenarios (sequences of events) that can lead to undesirable consequences. However, there is no guarantee that this process can ever be complete and that all possible scenarios have been identified and properly assessed. This lack of completeness introduces an uncertainty into the results and conclusions of the analysis.
(2) Modelling inadequacy. Even for those scenarios that have been identified, the event sequence and system logic models may not precisely represent reality. Uncertainties are introduced by the relative inadequacy of the conceptual models, the mathematical models, the numerical approximations, the coding errors and the computational limits. These uncertainties are discussed as part of the uncertainty analysis in the PSA, and sensitivity studies are usually performed to assess their relative importance.
(3) Input parameter uncertainties. These are uncertainties introduced as a result of scarcity or lack of data, variability within the populations of facilities or components, and assumptions made by experts. Input parameter uncertainties are most readily quantifiable.

Sensitivity studies are conducted to address the impact of selected modelling assumptions that could produce significant changes in the overall results. Sensitivity studies can be performed simply by replacing certain assumptions that can be quantified with others more or less conservatively and re-evaluating the model to compare results.

Importance measures are to be calculated and used to interpret the results of a PSA. The following importance values are typically used in Level 1 PSA:

(a) Fussell–Vesely importance;
(b) Risk reduction worth;
(c) Risk achievement worth;
(d) Birnbaum importance.

A discussion on uncertainty, sensitivity and importance analyses is provided in several IAEA publications applicable to Level 1 PSA (e.g. Ref. [7]). The treatment of uncertainties and the development of sensitivity analyses are further discussed in Section 3.4.12 as they apply to Level 2 PSA.

3.3.11. Documentation

It is common practice among PSA practitioners to present the results in a report that contains the following components:

(a) Executive summary;
(b) Main report;
(c) Appendices and annexes.

A typical table of contents for the Level 1 PSA report is presented in Appendix I.

3.4. MAJOR TASKS IN LEVEL 2 PROBABILISTIC SAFETY ASSESSMENT

3.4.1. Background

In Level 2 PSA, the unsuccessful sequences (typically core damage sequences) identified in Level 1 PSA are analysed from the point of view of both the behaviour of the core and the release of radioactive material from the damaged fuel to the environment. In this analysis, the responses of both the core and the confinement system are analysed following a sequence of events. This analysis starts with the end states defined in Level 1 PSA and ends with potential consequences for facility personnel or the public. In a typical Level 2 PSA, the end states are defined as a set of release categories.

The project management and organizational aspects of Level 2 PSA, including the definition of the objectives and the scope, are similar to those of Level 1 PSA, but the level of detail and the depth of the analyses involved depend on the power level and the complexity of the design of the research reactor, using a graded approach as specified in IAEA Safety Standards Series No. SSG-22, Use of a Graded Approach in the Application of the Safety Requirements for Research Reactors [28]. A critical facility with low hazard potential, passive systems, a small source term and no emergency confinement mode has a different Level 2 PSA than a reactor with a power level of several tens of megawatts and a containment or active confinement system. This subsection presents a generic approach to Level 2 PSA for a research reactor of low to moderate power. High power research reactors or research reactors with a complex design need to follow the guidance applied to nuclear power plants in SSG-4 [10].

3.4.2. Identification of relevant facility design features

The PSA team has to familiarize itself with the features of the facility design that can influence the progression of a severe accident. This exercise could include the reactor building as well as all defence in depth Level 4 systems inside and outside the building. For existing facilities, familiarization includes a facility walk-through with the participation of operating personnel.

Examples of the features that need to be identified (as applicable) are as follows:

(a) The behaviour of the core structures, fuel cladding and reactor pool boundary under unsuccessful sequences identified in Level 1 PSA.
(b) The configuration and behaviour of the confinement or containment in terms of transport and diffusion of hydrogen; a highly compartmentalized confinement or containment configuration will limit the extent to which combustible gases mix and disperse in the confinement or containment atmosphere.
(c) Features that could lead to confinement or containment bypass sequences (e.g. leaks, confinement or containment failure, penetration of seals).
(d) Emergency mode for confinement or containment (e.g. release rate, filter performance, isolation, venting, energy removal).

The information relevant to Level 2 PSA quantitative data that is necessary to carry out a facility specific analysis needs to be collected and organized. The data necessary for the PSA depend on the scope of the analysis and on the facility specific computer model used to calculate accident progression. Detailed architectural and construction data for the confinement or containment structure are collected to develop facility specific models of confinement or containment performance if such calculations are required by the scope of the project. An example of key data for Level 2 PSA is provided in Table 1.

TABLE 1. EXAMPLES OF KEY FACILITY AND CONFINEMENT OR CONTAINMENT DESIGN FEATURES TO BE CONSIDERED AS INPUT FOR LEVEL 2 PSA MODELLING

Key facility or confinement/containment design feature	Definitions and comments
Reactor:	
Reactor type	Open pool, tank in pool, pressurized research reactor
Power level	Total thermal power at steady state; power peak value and duration for reactors with pulse mode
Type of fuel meat/ fuel geometry/cladding	Metallic uranium, U_3O_8, U_3Si, UO_2–plate, pin–aluminium, stainless steel, zircaloy
Core:	
Mass of fuel and mass of cladding	Actual values
Fuel assembly geometry	Actual values
Type and mass of control rods/plates	Actual values
Spatial distribution of reactor power (neutron flux)	Axial and radial peaking factors
Decay heat	Decay heat as a function of time after shutdown
Radioactive material inventory	Inventory of radionuclides in the core; specify irradiation time
Reactor coolant system:	
Reactor coolant and moderator types	Light water, heavy water, other
Reactor coolant system volume	As designed and built
Pressure relief capacity	In case of potential for pressure buildup (valid only for specific type/high power research reactor)
Isolation of confinement/ containment penetrations connected to coolant system	Potential for confinement/containment bypass, if applicable

TABLE 1. EXAMPLES OF KEY FACILITY AND CONFINEMENT OR
CONTAINMENT DESIGN FEATURES TO BE CONSIDERED AS INPUT
FOR LEVEL 2 PSA MODELLING (cont.)

Key facility or confinement/containment design feature	Definitions and comments
Confinement or containment:	
Geometry	Shape and separation of internal volumes
Free volume	As built information
Design pressure and temperature	Realistic assessment of maximum values is required for the PSA
Material	Concrete, other
Operating pressure and temperature	Actual operation values
Hydrogen control mechanisms	Recombiners, igniters, inertness (if applicable; not applicable for designs in which there are no sources of hydrogen, such as reactors with MTR type aluminium clad fuel plate)
Concrete aggregate	Specify chemical content
Proximity of confinement/ containment boundary	Distance from reactor pool boundary
Response to external hazards	Structural damage due to external events applicable to the site
Potential for confinement reconfiguration/containment isolation failure	Reliability of seal materials, active components for reconfiguration/isolation
Potential for cooling of deformed/molten core	If applicable

Note: MTR — materials testing reactor.

3.4.3. Definition of facility damage states

Facility damage states can be classified into two main classes as specified in SSG-4 [10]:

(1) Those in which radioactive material is released from the reactor coolant system to the confinement or containment and the confinement or containment remains intact and reconfigured to emergency mode/isolated;

(2) Those in which radioactive material is released from the reactor coolant system to the confinement or containment, but the confinement or containment is either bypassed or ineffective.

For case (1), a confinement/containment event tree (CET) analysis can be performed. For case (2), source term analysis (see Section 3.4.11) only may suffice, although the CET may be needed to take into account possible facility features that can reduce the source term (e.g. filtered releases versus unfiltered releases). Some examples of the attributes that may need to be accounted for in defining facility damage states are given in Table 2.

TABLE 2. TYPICAL ATTRIBUTES FOR DEFINING FACILITY DAMAGE STATES

Status	Attribute
Initiating event	— Large LOCA — Small LOCA — Beam LOCA
Status of reactor coolant system at time of core damage	— Location of break — Forced convection through core — Natural convection through core — Drained
Status of emergency cooling/emergency water make-up	— Injection failure — Available
Status of confinement/containment	— Reconfigured/isolated/failed — Venting actuated/failed — Filtered recirculation actuated/failed

Note: LOCA — loss of coolant accident.

For determination of the source term, as specified in SSG-22 [28], and the release from the core to the confinement or containment, it is necessary to analyse the extent of the core damage early.

In research reactors, where aluminium is a major component in the fuel, the core damage may be defined as one of the following (in order of increasing conservatism):

(a)   The temperature of the cladding rises above the blistering temperature;
(b)   The ratio between the critical heat flux and the maximum heat flux (average heat flux × radial power peaking factor × axial peaking factor) falls below the acceptance criteria (typically a value greater than 1.3 (up to 2) to account for uncertainties in the experimental correlations for critical heat flux);
(c)   The ratio between the critical heat flux to cause onset of significant void and the maximum heat flux (average heat flux × radial power peaking factor × axial peaking factor) falls below the acceptance criteria (typically a value greater than 1.3 (up to 2) to account for uncertainties in the experimental correlations for onset of significant void);
(d)   The temperature of the cladding rises above the temperature of onset of nucleate boiling (for low power research reactors and critical facilities, it is often considered as a conservative safety criterion).

In many cases, Level 2 PSA for a research reactor assumes 100% core damage. This may be possible for low power, limited inventory cores. For more complex and higher power reactors, the extent of core damage has to be determined.

There are currently no computational codes that allow a detailed analysis of the progression of the core following damage in materials testing reactor type fuel cores. For this type of fuel, there are no hydrogen related phenomena. In the case of research reactors with zircaloy or stainless steel clad fuel (e.g. TRIGA), the type of phenomena that need to be analysed will depend on the amount of energy released into the fuel and the temperatures reached by the cladding. For zircaloy or stainless steel clad fuel, a similar methodology to that used for nuclear power plants can be adopted. For other types of fuel, the extent of damage needs to be determined with an estimative approach.

The applicability of codes for severe accident modelling for nuclear power plants has to be evaluated before they are used for Level 2 PSA for a research reactor, as certain codes have strong phenomenological approaches.

Once the accident sequences have been analysed and the amount of energy released into the fuel has been calculated, the sequences can be grouped using this parameter as the leading value: damage states will be determined by the amount of energy that is released into the fuel. The temperature of the fuel following this deposit of energy can be evaluated through analytical means and the final extent of damage can be estimated.

After the extent of damage has been determined, the amount of radioactive material released has to be calculated. The fraction of damaged core determines the material available for release. The amount of material released into the primary water will be given by the retention factors adopted for the fuel. Retention factors have been determined experimentally. An example of values of retention factors can be taken from appendix B of Ref. [6]. The following are typical values of retention factors for melted fuel (F_1):

— F_1 (noble gases) = 1;
— F_1 (iodine, tellurium, caesium) = 0.27;
— F_1 (barium, strontium, ruthenium) = 0.03;
— F_1 (fluorine, phosphorus, others) = 0.001.

Following release from the fuel, the retention factors for primary water are applied to obtain a release to the reactor hall. This is the inventory that will be available for Level 2 and 3 PSA calculations. Typical values of the retention factors for primary water can be found in appendix B of Ref. [6], and are presented in Table 3.

3.4.4. Facility damage states not initiated by bypass of the confinement or containment

The equipment and system failures identified in Level 1 PSA that could either affect or challenge the confinement or containment or the release of radioactive material have to be taken into consideration to determine the facility

TABLE 3. TYPICAL VALUES OF RETENTION FACTORS FOR PRIMARY WATER [6]

	Without expulsion of water	With expulsion of water	Fusion in air
F_2 (noble gases)	2×10^{-2}	4×10^{-1}	1
F_2 (iodine)	1×10^{-4}	5×10^{-3}	1
F_2 (solids)	1×10^{-4}	1×10^{-4}	1

damage states that are not initiated by bypass of the confinement or containment. Furthermore, the following aspects also have to be taken into account:

(a) Initiating events that may affect the progression of the core melt and hydrogen generation, the rate of release of radioactive material into the confinement or containment, and the time interval of the release of radioactive material;
(b) Failure mode of the core cooling function, which can lead to core meltdown;
(c) The quantification of damage suffered by fuel.

The integrity of the confinement or containment and the functioning of associated engineered safety features play a vital role in determining the response of the confinement or containment and have to be taken into account in the grouping of accident sequences into facility damage states. They may also influence the air circulation of the confinement or containment, the mixing of radioactive and combustible gases, and a reduction in airborne activity.

3.4.5. Facility damage states with bypass of the confinement or containment

It is necessary to identify the attributes with a bearing on the attenuation of the concentrations of radioactive material along the release pathway. The timing of the attenuation may also affect the facility damage states with bypass of the confinement or containment. This has to include failure time and status of the emergency core cooling system and emergency water make-up system. Whether the leak pathway is isolatable, or whether it passes through water, has to be considered. The status of the emergency exhaust filtration system, heating, ventilation and air-conditioning could have a significant impact on the facility damage states and has to be taken into account.

3.4.6. Final selection of facility damage states

Most research reactors have a small number of final facility damage states. However, for more complex reactors with higher power or many experimental facilities that contribute to the source term, the consideration of all factors and parameters that affect Level 2 PSA may result in a large number of potential facility damage states. These can be reduced to a manageable number using two approaches. One approach is to combine similar facility damage states and perform a conservative analysis to select a representative sequence that characterizes the facility damage state for the purpose of Level 2 PSA. The second approach may involve the use of a frequency cut-off and the screening out of less important damage states.

3.4.7. Data compilation and analysis

Data need to be obtained from valid sources, such as the following:

(a) Design documents or facility licensing documents (e.g. safety analysis report);
(b) As built drawings;
(c) Facility specific operating, maintenance or test procedures;
(d) Engineering calculations or analysis reports;
(e) Observations during facility walk-throughs;
(f) Construction reports;
(g) Vendor manuals.

The data sources have to be recorded in the PSA report.

3.4.8. Confinement or containment analysis

Once the release of radioactive material from the fuel and the water has been established, the behaviour of the confinement or containment and the progression of the accident are analysed in order to determine the release from the confinement or containment. For low power, low source term research reactors (e.g. power ≤2 MW), it may not be necessary to model the behaviour of the confinement system. For higher power levels and more complex research reactors, the analysis is similar to the approach presented for nuclear power plants in SSG-4 [10].

Detailed information on the structural design of the confinement or containment and its penetration has to be collected to accurately assess the performance of the confinement or containment. Information on the potential for leakage through seals or penetrations is to be given special consideration. Typical information needed to assess the performance of confinement or containment is provided in Table 4.

TABLE 4. TYPICAL INFORMATION NEEDED FOR CONFINEMENT/CONTAINMENT PERFORMANCE ASSESSMENT

Information type	Necessary detail
Confinement/containment material	Steel
	Concrete
	— Pre-stressed
	— Post-tensioned
	— Reinforced

TABLE 4. TYPICAL INFORMATION NEEDED FOR CONFINEMENT/
CONTAINMENT PERFORMANCE ASSESSMENT (cont.)

Information type	Necessary detail
Seals/penetrations	Equipment hatches
	Personnel hatches
	Piping penetrations
	Purge and venting lines
	Cable penetrations
Other	Geometrical shape
	Geometrical discontinuities
	Liner or no liner
	Interactions with surrounding structures

Depending on the scope of Level 2 PSA, existing calculations for facilities with similar designs could be used. In this case, the PSA report has to provide a justification for the use of existing calculations by demonstrating the similarities of the designs and the applicability of the existing structural response analyses to the facility under consideration.

Consideration has to be given to various types of load on the confinement or containment (e.g. static pressure loads, localized heat loads, localized dynamic pressure loads). The supporting analyses provide an engineering basis for confinement or containment failure mode, location, size and ultimate pressure and temperature capabilities.

A CET has to be developed for each damage state to model the behaviour of the confinement or containment system. The methodology for the development of the CET can be summarized as follows:

(a)   Identification of the components of the confinement or containment system that will intervene in the accident sequence;
(b)   Identification of the passive components of the confinement or containment system that may fail during the accident sequence;
(c)   Identification of operators' actions that may be needed for accident recovery;
(d)   Elaboration of event trees.

A typical approach for developing a CET for Level 2 PSA is presented in Table 5.

Attention has to be paid to the environmental conditions, including temperature, humidity and radiation field for the systems that will actuate in the time frame described in Level 2 PSA.

TABLE 5. TYPICAL APPROACH FOR DEVELOPING A CONFINEMENT/
CONTAINMENT EVENT TREE FOR LEVEL 2 PSA

Top event question	Prior dependency	Question type
Phase 1: Initiating event through to early period of in-vessel core damage		
1. Is the confinement in emergency mode/ containment isolated?	No	Based on facility damage state
2. What is the status of the fans of the ventilation system?	No	Based on facility damage state
3. Does the reactor cooling system need to be depressurized? Has it been depressurized?	Yes	Emergency procedures
4. Does the confinement/containment fail in the very early time frame?	Yes	Accident progression
5. Is the confinement/containment safety function recovered in the very early time frame?	Yes	Based on facility damage state
Phase 2: Late period of damage progression. Includes breach of reactor tank in pool or tank pressurized designs		
6. Is core damage limited to cladding failure, with no core melt?	Yes	Accident progression
7. Does the core remain underwater?	Yes	Accident progression
8. Is core damage arrested inside the tank? Is there core material exiting the reactor vessel/pool?	Yes	Accident progression
9. Is there exothermic fuel/coolant interaction?	Yes	Accident progression
Phase 3: Long term response of the facility		
10. Is there AC power?	Yes	Based on facility damage state

TABLE 5. TYPICAL APPROACH FOR DEVELOPING A CONFINEMENT/
CONTAINMENT EVENT TREE FOR LEVEL 2 PSA (cont.)

Top event question	Prior dependency	Question type
11. Does the emergency cooling system actuate?	Yes	Based on facility damage state/ accident progression
12. Is core material in a coolable condition outside the vessel/tank?	Yes	Accident progression
13. Does confinement/containment fail in the late time frame?	Yes	Accident progression
14. What are the modes of confinement/ containment failure?	Yes	Accident progression

Level 2 PSA ends with a calculation of the inventory released through the stack, venting system or building (in the case of failure of the confinement or containment). Releases through the stack and venting system have to take into account the retention factors of the appropriate filters.

3.4.9. Characterizing end states of confinement/containment event trees

Confinement or containment analysis results in spectrum end states. Each end state comprises the facility damage state, the associated state and the associated sequence of events, which might include an event associated with containment system success or failure. The end state represents in-core and out-of-core physical phenomena and containment system performance, which determines the amount and level of release from the containment. The following factors characterize the release:

(a) The failure mode of the reactor coolant system;
(b) The mode and time of confinement or containment failure;
(c) The alternative or emergency cooling mode for reactor core and associated structural systems;
(d) The cooling mechanism as part of severe accident management;
(e) The success and failure of radioactive material retention in containment.

End states with similar characteristics, physical phenomena, containment and mitigation system responses, and time and location of release, as given in Table 6, are grouped together. These groups are referred to as release categories, meaning that the number of deterministic analyses required to assess the containment source can be reduced to a manageable figure.

3.4.10. Grouping end states of confinement/containment event trees into release categories

The end states of the CET are grouped into the specified release categories. The grouping of the end states of the CET has to be carried out with regard to the various factors that affect the release of radioactive material. All the end states of the CET, within a particular set, with similar release characteristics and off-site consequences, need to be grouped together. Their frequencies of

TABLE 6. TYPICAL ATTRIBUTES USED TO SPECIFY RELEASE CATEGORIES FOR LEVEL 2 PSA

Release attribute	Variation
Time frame in which release begins	— At the onset of core damage (e.g. bypass of confinement/containment) — Early (during in-pool core damage) — Late (e.g. after core is uncovered or after core has breached the reactor vessel)
Modes or mechanisms of confinement/containment failure or leakage	— Design basis failure/leakage — Design extension condition failure/leakage — Catastrophic
Active systems that capture radioactive material	— Sprays — Filters
Location of release	— Ground level — Elevated
Energy of release	— Minimal — Highly buoyant
Release rate	— Quick, puff — Slow, continuous — Multiple plumes

release categories are then added together to obtain the overall frequency of release categories.

3.4.11. Source term analysis

For low power and low source term research reactors, conservative assumptions may be made about the source term to simplify analysis and demonstrate that, even with large uncertainties caused by the lack of specific modelling for the facility, compliance with acceptance criteria can be demonstrated, as described in SSG-22 [28]. The design features of a research reactor and its accident characteristics influence the magnitude of the source term and its release characteristics (e.g. the composition and configuration of the fuel assembly and the control assembly, the magnitude and variation in power density within the core, the magnitude and variation in fuel burnup, and the composition of the core structural material and shielding). Typically, the following factors need to be considered in source term analysis:

(a) The coolant inventory during core damage and at the time of release of material from the pool;
(b) The availability of coolant;
(c) The amount, depth and composition of core debris;
(d) The mode of operation of the confinement or containment safety equipment (e.g. sprays, recombiners);
(e) The size of the confinement or containment breach (i.e. leakage rate);
(f) The location of the failure and the resulting transport pathway to the environment.

Source term analysis can be facility specific or can be adopted from a similar facility. The facility specific analysis necessitates modelling of all the processes that affect the release and transport of radioactive material inside the confinement or containment and in adjacent buildings, including the following:

(a) Radioactive material release from the fuel during core damage;
(b) Retention of radioactive material within the primary coolant system;
(c) Release of radioactive material during the uncovering of the core;
(d) Retention of radioactive material inside the confinement or containment (e.g. due to iodine plate-out).

The spatial distribution of the radionuclide concentrations within the reactor coolant circuit and the confinement or containment, as well as the quantity released into the environment, have to be estimated as a part of the source term analysis. The analysis has to be carried out for a sufficient number of accident sequences in each release category, as specified in SSG-4 [10]. The source term analysis for a relatively small number of accident sequences can be considered as acceptable, provided that each release category contains similar accident sequences and the phenomena that drive the release have a relatively low uncertainty. In cases in which the release is driven by energetic phenomena such as direct heating owing to fire, or involves phenomena that have a relatively high level of uncertainty, source term analysis has to be carried out for a number of accident sequences to provide confidence that the source term has been characterized accurately.

3.4.12. Uncertainties

Issues giving rise to uncertainties in source terms include the following:

(a) Uncertainties in processes leading to core damage;
(b) Uncertainties in the behaviour of confinement or containment;
(c) The effect of fuel burnup on the release rate of radioactive material from fuel;
(d) The chemical forms of volatile and semi-volatile radioactive material;
(e) The chemical reaction of coolant and/or gases with cladding, fuel, neutron absorbers and structural materials;
(f) The deposition rates of radioactive material and aerosols on the surfaces of the reactor coolant circuit;
(g) The release of radioactive material and aerosols during molten core–concrete interaction (unlikely for low power research reactors);
(h) The chemical processes during molten core–concrete interaction (unlikely for low power research reactors);
(i) The chemistry of radioactive material captured in the coolant and water pool;
(j) Revaporization and resuspension of radioactive material from surfaces;
(k) The chemical decomposition of radioactive material aerosols.

The uncertainties in the source term quantification could be addressed by carrying out sensitivity studies for the major sources of uncertainty that influence the results of Level 2 PSA.

3.4.13. Human reliability analysis

Human reliability analysis in Level 2 PSA follows the same methodology adopted for Level 1 PSA in Section 3.3.5.2, with special consideration for errors of commission and omission in facility recovery and accident management actions.

3.4.14. Documentation

Documentation for a Level 2 PSA has to provide sufficient information to satisfy the objectives of the study and to facilitate its subsequent refinement, update and maintenance in the light of changes to facility configuration or technical advances in the methodology for these analyses. In addition, the Level 2 PSA report needs to be easily updatable to maintain a living PSA concept.

The Level 2 PSA report has to clearly document the findings of Level 2 PSA, including the following:

(a)   Facility specific design or operational vulnerabilities;
(b)   Key operator actions for mitigating severe accidents;
(c)   Potential benefits of various engineered safety systems.

The information concerning the organization of documentation provided for Level 1 PSAs also applies to Level 2 PSAs. The report for Level 2 PSAs can be divided into three major parts:

(1)   Executive summary;
(2)   Main report;
(3)   Appendices.

Conclusions are to be unambiguous, reflecting the main generic results drawn from the analysis of uncertainties associated with phenomena, models and databases, and the contributory analyses. The effect of the underlying assumptions, uncertainties and conservatisms in the analyses and methods on the results of Level 2 PSA has to be clearly reflected through the sensitivity studies.

3.5. MAJOR TASKS IN LEVEL 3 PROBABILISTIC SAFETY ASSESSMENT

3.5.1. Background

Level 3 PSA[5] provides a quantification of off-site radiological consequences of accidents, which in general is based on the results of Level 2 PSA. From these insights, the assessment evaluates the relative effectiveness of emergency response planning aspects of off-site accident management, and the economic impacts.

The objectives of Level 3 PSA include probabilistic estimates of health, environmental and economic consequences of accidents, thus allowing for a more complete understanding of the risk associated with a research reactor. The Level 3 assessment can be performed with a relatively low effort when sufficient information is available on the spectrum of radioactive releases to the environment and the characteristics of that environment, and the people living within it.

The scope of Level 3 PSA includes the consideration of all the radionuclides that might be released during an accident, as identified in Level 2 PSA, and their potential dispersion in the surrounding environment. As such, the scope of Level 3 PSA needs to be determined in accordance with the surrounding environment. The radionuclide release can be both into the atmosphere (and eventually into the soil) and into water bodies. The most commonly used Level 3 related safety criteria are: (i) the individual risks of early and late fatal health effects; (ii) the societal risk of early and late fatal health effects; and (iii) the risks of unacceptable land contamination. In some cases, a licensee may be required to characterize and reduce off-site risks.

Compared with the information from Level 1 and Level 2 PSAs, the information obtained in Level 3 PSA allows for a better and far more complete assessment and characterization of the off-site (public) risks attributable to a spectrum of possible accident scenarios involving a research reactor. Level 3 PSA can thus also be used in the analysis of the environmental impact statement. In this case, its results can be used in the decision making process for a new research reactor project.

---

[5] At present, there is no IAEA publication covering the area of Level 3 PSA. A new IAEA TECDOC is at an advanced stage of development and will provide a comprehensive discussion of Level 3 PSA. Section 3.5 of this Safety Report refers to selected technical content from this future IAEA publication.

3.5.2. Description of the radionuclide release

The starting point for a radiological consequence assessment is the release of the radionuclides into the environment beyond the engineered structures, which is most commonly derived from the Level 2 source term analysis. Information on this radionuclide release, which is referred to as the 'accident source term', is provided for each of the release categories to be included in the assessment. The accident source term describes the characteristics of the release of radioactivity. These characteristics may include the following:

(a) A specification of the initial inventory available for release;
(b) The total amount of releases, including release fractions and their time dependence;
(c) The approximate location of the release relative to the large structures on-site (e.g. isolated or entrained in the wakes of the buildings and whether the energy in the release is sufficient to significantly increase the effective release height);
(d) The physical and chemical form of the release, particularly the assumptions about iodine, and particulate/aerosol size distribution.

The release fraction describes the fraction of the available inventory that is released to the environment. The product of the release fraction and the inventory available for release (measured in Becquerel) gives the quantity of each radionuclide that is released. To simplify calculations for the source term analysis, the radionuclides are usually grouped according to their chemical and physical properties, such as noble gases, halogens, alkali metals, tellurium group, noble metals, lanthanides and cerium group [29][6].

3.5.3. Environmental transport mechanisms

There are three main pathways through which people can be exposed to radiation from nuclear activities: (i) atmospheric dispersion and deposition; (ii) dispersion and deposition via surface water and groundwater; and (iii) at a short distance from the source by direct irradiation. The releases to the atmosphere through dispersion and deposition are the principal concern and, thus, need to be carefully assessed.

Radionuclides released to the atmosphere owing to a severe nuclear accident, primarily as fine aerosol (usually with a mass median aerodynamic

---

[6] Table 5 of Ref. [29] lists the elements in each radionuclide group to be considered in the source term analysis.

diameter of 1–5 μm) but partially as gas, create a plume that is carried downwind. Under this transport mechanism, the plume expands horizontally (cross wind) and vertically owing to molecular diffusion and turbulent eddies in the atmosphere. The two processes (i.e. diffusive and turbulent mass transport) are collectively referred to as 'dispersion'. The results for the primary atmospheric transport needed in a consequence analysis are the transient air and ground concentrations at each location affected by the plume. The transient nature of the transport processes is needed to quantify doses to populations that evacuate and relocate.

Subsequent to the release, the radionuclide content diminishes, both by radioactive decay and through deposition mechanisms. The latter can be further divided into two categories: 'dry' and 'wet'. Dry deposition is practically a surface effect, whereby material in contact with the ground is removed through a number of processes, including gravitational settling, impaction onto surface projections, molecular diffusion of gases and Brownian diffusion of particles. The velocity of dry deposition of a specific radionuclide will depend inter alia on its chemical form, its particle size, the atmospheric conditions (e.g. temperature, humidity, wind speed, atmospheric stability class), and the nature of the surface onto which it is depositing [30]. Wet deposition is the removal of released material as a result of either the interaction with falling precipitation (washout) or the incorporation of the contaminant into rain clouds that create precipitation (rainout).

3.5.4. Exposure pathways and dose calculation

There are eight exposure pathways by which people can accumulate radiation doses after accidental releases of radioactive material to the atmosphere [30, 31]:

(1) External irradiation (beta, gamma) from radionuclides in the passing plume or cloud, referred to as 'cloud shine';
(2) External irradiation (beta, gamma) from radionuclides deposited on the ground, referred to as 'ground shine', and on other surfaces;
(3) External irradiation from radionuclides deposited on the skin and clothing;
(4) External exposure due to direct irradiation from the source;
(5) Internal irradiation from radionuclides inhaled directly from the passing plume;
(6) Internal irradiation from radionuclides inhaled following resuspension of deposited material;
(7) Intakes of radionuclides due to the inadvertent ingestion of radioactive material deposited on the ground or on other surfaces;
(8) Intakes of radionuclides due to the consumption of contaminated food and water.

The source terms determine the relevant exposure pathways. For example, a noble gas release will lead to doses caused by direct irradiation from the passing plume, whereas a release containing actinides will create doses through inhalation and ingestion.

The doses from all important radionuclides and from all relevant exposure pathways need to be added together to obtain the effective dose for the whole body. The important radionuclides may include those that significantly contribute to radiation doses, accounting for radionuclide inventory, release fractions, radioactive decay prior to exposure and the relative dose per activity released. Doses from internal exposure pathways are commonly integrated over a person's lifetime to account for the residence time of the radionuclide in the human body; these doses are referred to as 'committed doses from intakes'. Dose coefficients for the internal exposure pathways account for radioactive decay, production of decay products and their contributions, and metabolic processing of the radionuclide by the human body. Doses from external exposure pathways are received only when the person continues to be exposed to the source of radiation. Accordingly, a time period will need to be assumed and the dose needs to be integrated over this period.

3.5.5. Economic consequences

Many of the consequences of an accidental release of radionuclides can be translated into economic consequences. This provides a measure of the impact of the accident and enables the different effects of the accident to be expressed in the same terms and combined as appropriate. Information on economic activity, such as gross domestic product by economic sector, or land and housing values, is usually available on a regional basis and sometimes on a country basis. The specific economic model may be used to determine the types of economic data needed.

3.6. LOW POWER AND SHUTDOWN PROBABILISTIC SAFETY ASSESSMENT

3.6.1. Background

Level 1 PSA for low power and shutdown modes (LPSD PSA Level 1) is performed to assess the risk contribution from low power and shutdown states of the reactor. Many factors contribute to this risk, including but not limited to the following:

(a) The facility and system configuration changes could make management of the safety function more complex.
(b) In research reactors, core configuration changes have the potential to lead to reactivity related incidents.
(c) Items important to safety may be temporarily unavailable because of maintenance and surveillance activities.
(d) The number of activities that may stress communications with the control room or even between operating teams is considerable.
(e) The human factor is crucial as procedures are sometimes carried out simultaneously.
(f) Refuelling operations are performed manually in most research reactors.
(g) In low power operation, the power trip margins can be very large, potentially leading to steep power increase gradients (a form of startup accident).

Therefore, LPSD PSA Level 1 needs to form part of a full power Level 1 PSA. The objective of LPSD PSA Level 1 is to assess the potential for core damage or fuel damage involving scenarios and conditions in the low power and shutdown modes of the facility, and to obtain risk insight into the design and operation of the research reactor during such modes, including the following:

(a) Reactor startup and operation at low power;
(b) Reactor shutdown;
(c) Maintenance and outage management, including refuelling operations;
(d) Experimental and engineering test loop operations;
(e) Fuel transfer and storage.

The results of LPSD PSA Level 1 are as follows:

(a) Assessment of CDF;
(b) Individual fuel damage;
(c) Insights into facility design and operational weaknesses;
(d) Consequences of experimental facility accidents.

The insights obtained from LPSD PSA can be used to improve overall outage management plans, facility startup, facility shutdown and test and maintenance procedures, as well as to identify human actions that could affect safety.

The scope of LPSD PSA Level 1 includes the following specific aspects:

(a) Transient condition from high power to low power, low power to shutdown, fuelling operation, various facility test and maintenance configurations, reactor startup to low power, and power increase from the low power to the full power state;
(b) Consideration of external and internal hazards;
(c) Sources of radiation, such as the reactor core, storage and experimental facilities;
(d) Consideration of the human factor, uncertainty and sensitivity, and importance analysis;
(e) Targeted application, if any, such as risk based maintenance management and in-service inspection programmes.

3.6.2. Major steps

The full power Level 1 PSA model forms the reference model for the initiation of LPSD PSA Level 1. The initial Level 1 PSA task involves revisiting the list of initiating events, system models and data to identify the scenarios for LPSD PSA Level 1. The major steps in LPSD PSA Level 1 are similar to those in the full power Level 1 PSA, and include the following:

(a) Selection of initiating events as applicable to the low power and shutdown states of the facility;
(b) Identification of FOSs, taking into account the specific features of the facility;
(c) Event sequence modelling considering the availability of the safety function as applicable to each FOS;
(d) Accident sequence modelling;
(e) System modelling by revisiting the reference fault tree models in the full power state;
(f) Reliability data collection and analysis specific to the shutdown state;
(g) Human factor assessment, identification of human actions leading to accident initiators and post-accident human actions that may recover or aggravate the evolving accident condition;
(h) Accident sequence quantification;
(i) Uncertainty, sensitivity and importance analysis;
(j) Assessment of potential risk from:
    (i) Experimental and engineering test loops;
    (ii) Refuelling induced LOCA or reactivity transients;
    (iii) Ex-core phenomena such as fuel handling and storage operations.

Uncertainty and sensitivity analyses, along with the assessment of human factors, also form part of LPSD PSA. In fact, the human factor assessment is more complex in LPSD PSA than in full power PSA because of the greater manual and procedural content of operational activities during this time.

Research reactor operations are also characterized by an availability factor that is typically lower compared with nuclear power plants. Research reactors require frequent shutdowns owing to scheduled shorter cycle lengths (e.g. owing to isotope assembly handling, experimental requirements and lattice experiments in support of reactor physics studies). In fact, many facilities only operate for one or two shifts at a time. This aspect is taken into account to assess the annual frequency of initiating events for the shutdown state.

3.6.2.1. *Project organization and management*

Before initiating LPSD PSA Level 1, a project report is prepared that includes the objective, scope, resource requirements (e.g. human resources, funding, logistic support), organizational aspects (e.g. technical collaboration with other organizations or departments in the same institute), tentative schedule (e.g. bar chart listing target time required for various activities), peer review requirements, deliverables (e.g. executive summary, main report that provides recommendations for reducing shutdown risks) and quality assurance procedures. The project report serves as a link between the PSA team, management and other departments and allows for a better appreciation of requirements related to the project and its deliverables.

The following major inputs are required for LPSD PSA:

(a) A list of the test and surveillance programmes, interdependencies and specific conditions relevant to the assessment of low power or shutdown risks;
(b) Level 1 PSA for the full power condition as a reference model;
(c) A safety analysis report to provide deterministic insights into accident conditions;
(d) A reliability database for SSCs specific to repair and maintenance activities;
(e) Facility operating and emergency operating procedures to assess human interactions that can either result in an initiating event or adversely affect normalization procedures;
(f) An incident database to determine the precursor for an initiating event;
(g) Design and operating manuals for SSCs;
(h) Training documents;
(i) A record of maintenance and surveillance activities during shutdown state for at least five years;

(j)   Facility maintenance records on equipment;

(k)   Facility reports and logs.

A quality assurance programme has to be developed for various activities, such as initiating event selection, accident sequence modelling, system modelling, data collection and analysis, accident sequence evaluation, interpretation of results, formulation of recommendations and documentation. The quality assurance programme also addresses activities related to final documentation, and aspects related to detailed data collection, data analysis, major and system specific assumptions, and failure and success criteria. It is recommended that during a PSA project, reports are issued at various stages, such as systems analysis reports issued for internal reviews. These reports have to be accompanied by a quality assurance checklist that addresses major aspects of the analysis, such as software used, input data (facility specific or generic), considerations related to human factors, approach and models used for human reliability analysis, whether uncertainty analysis is carried out, and validation of assumptions by sensitivity analysis. Such a checklist provides the first level of confidence in the PSA study.

3.6.2.2.   *Identification of facility operational states*

During the low power and shutdown states, the reactor can have different modes, depending on the status of various systems. However, for the purpose of LPSD PSA Level 1, broad categories of reactor states have to be formulated. The reactor states requiring similar considerations can be grouped into one representative or enveloping state. One such FOS could be reactor refuelling in progress, with the reactor decay heat removal system operating and the primary coolant system in the shutdown state, safety rods in a poised state such that any reactivity excursion in the shutdown state can be detected and detained, and the containment integrity maintained.

It is important to recognize the potential risk contributions from experimental facilities and core configuration changes applicable to low power or shutdown states of the reactor. For example, the reactor may remain in the shutdown state while an engineering test loop may be operating under high temperature and high pressure conditions. Such conditions have to be considered when formulating the FOSs. Any activity that is related to core reactivity considerations, such as the calibration of neutronic channels, detector testing, replacement of control or shut-off rods, fuelling or isotope handling, has to be evaluated for risk potential.

An example of a list of FOSs is given in Appendix II.

3.6.2.3.   Identification of initiating events and screening

In most cases, LPSD PSA Level 1 addresses the following set of events:

(a)   Power supply failure to facility and equipment;
(b)   Events or conditions that may pose a threat to the decay heat removal function;
(c)   Actions or tasks that may pose a threat to the integrity of the primary cooling system;
(d)   Actions or tasks that have the potential for loss of coolant;
(e)   Scenarios adversely affecting core reactivity control;
(f)   Material handling in general and dropping of heavy loads in particular that can damage the fuel.

These categories of events have the potential to damage the reactor core. These categorizations are based on the requirements for similar safety functions, their success criteria and recovery mechanisms that include human intervention for recovery.

Compared with a full power operating state PSA, LPSD PSA requires modelling of aspects such as the identification of initiating events that are specific to shutdown tasks such as refuelling, in which the reactor core or coolant system boundary may not remain the same as in the full power operating state. The risk considerations from, for example, refuelling operations, require consideration of initiating event analysis such as refuelling induced LOCA or reactivity transients during refuelling, and initiating events outside the reactor core, such as ejection of fuel, failure of fuel cooling during fuel handling operations and fire that could damage fuel in the containment or confinement building. This means that, apart from the CDF matrix, an additional matrix related to damage to the single fuel assembly needs to be determined.

While a list of all possible or probable initiating events is crucial for making LPSD PSA Level 1 as comprehensive as possible, the screening of the initiating events has to be carried out using a similar procedure to that applied in full power Level 1 PSA.

The following are typical examples of initiating events for low power and shutdown states for a reference research reactor:

(a)   Transients:
      (i)   Reactivity transients during low power operation;
      (ii)   Reactivity transients during the shutdown state due to core loading changes;
      (iii) Reactivity transients during fuel storage and handling;

      (iv)   Other transients.
(b)   Loss of off-site power.
(c)   Loss of coolant accident:
      (i)   Excessive leakage from main coolant pump mechanical seal;
      (ii)   Maintenance induced LOCA;
      (iii)   Fuelling induced LOCA;
      (iv)   Heat exchanger tube failure.
(d)   Failure of cooling during fuel handling operation.
(e)   Loss of ultimate heat sink.
(f)   Loss of residual heat removal system.
(g)   Heavy load drop incidents.
(h)   Internal flooding.
(i)   Internal fire.

3.6.2.4. System modelling

This step requires revision of the full power Level 1 PSA fault trees. When modifying the fault trees, the FOS and initiating event being mapped have to be considered. Dependencies on the availability of safety systems during the postulated event have to be thoroughly analysed. The rest of the fault tree analysis procedure is the same as it is for the full power condition.

3.6.2.5. Data collection and analysis

This step is similar to that in full power Level 1 PSA, apart from the procedure for the assessment of initiating event frequency.

In the full power operating state, the initiating event frequencies are generally quantified based on an annual rate of occurrence. As the facility remains in the shutdown state for part of the time, this aspect needs to be factored in while defining initiating event frequencies. Furthermore, in shutdown PSAs, the initiating events have to take into account considerations such as equipment configurations, facility operational limits and conditions, system interlocks and outage management policy, which in turn are related to FOSs. The following examples show the assessment of the frequency of initiating events per calendar year (f_a).

(a)   Model I. The basic assumption in support of this model (1) is that the likelihood of a random event is proportional to the duration of time for which the event is being predicted.

$$f_a = f_h \times t_{fos} \qquad (1)$$

where

f_h is the hourly rate of occurrence of initiators in FOS;

and t_{fos} is the duration of FOS (hours in FOS per year).

For example, assuming that the facility specific class IV failure frequency during refuelling operations is 0.5/a (or 5.7×10^{-5}/h), and assuming further that the total fuelling time per year is 200 h, the annual frequency, f_a, will be $5.7 \times 10^{-5} \times 200 = 0.011$/a or 1.3×10^{-6}/h in that FOS. This model is applicable when data on event failure frequency during a given FOS are available.

(b)   Model II. This model (2) is used when the annual frequency of an initiating event is assessed not from the given occurrence data of the initiating event, but from the frequency of precursors to the initiating event. The input data required in this case are frequency of precursor per hour (f_{ph}) in a given FOS, conditional probability of an initiating event after the precursor has occurred, and duration of the FOS in hours.

$$f_a = f_{ph} \times P(\text{IE} \,|\, p) t_{fos} \qquad (2)$$

For example, voltage drop in class IV power supply to a defined level of, say, 85% of the full voltage could be taken as a precursor to power supply failure. Assuming the frequency of the voltage dipping to 85% or less occurs 40 times per year and the class IV failure frequency is 2/a, the conditional probability (P) of the initiating event (IE) corresponding to the precursor (p) is $2/40 = 0.05$. Furthermore, assuming that the f_{ph} value for the refuelling operation is 2×10^{-3}/h and the $t_{fos} = 600$ h/a, the annual frequency of class IV failure for a given FOS $= (2 \times 10^{-3}$/h) $\times 0.05 \times 600$ h/a $= 6 \times 10^{-2}$/a.

(c)   Model III. This model (3) enables an estimation of the initiating event frequency when considering the discrete number of entries of precursor for a FOS in a year and not the duration of a precursor (as was the case in the previous model).

$$f_a = n_{fos} \times f_{fos/a} \times P(\text{IE} \,|\, p) \qquad (3)$$

For example, the expected number of precursors from the facility records (voltage drop below 85%, n_{fos}) has a rated value of 0.75 (in fuelling operation), and the expected number of entries into the refuelling operation per year ($f_{fos/a}$) is 4, while $P(IE|p)$ has been estimated as 0.05. The annual frequency of the initiating event f_a can then be given as $0.75 \times 4 \times 0.05 = 1.5 \times 10^{-1}/a$.

3.6.2.6. Event sequence modelling and analysis

This step is similar to that in a full power Level 1 PSA, except that in this case the availability and capability of safety systems or recovery procedures have to be analysed to assess their ability to impede an evolving initiating event. In the shutdown state, some safety systems may either not be available or, if they are available, there may be a reduction in the level of redundancy or diversity. Human recovery actions will also be governed by the available time window before damage occurs. Furthermore, it may be possible to identify alternative means to prevent damage from occurring. For the most part, the event tree approach has to be employed for event sequence modelling.

The event trees developed for a full power operating state PSA need a careful review for their applicability to LPSD PSA Level 1. Accident sequences involving some initiating events may lead to single fuel damage (e.g. during refuelling or operations related to the test loop). Some of the characteristics of the facility shutdown state of the reactor include a reduction in the capability of safety systems (e.g. a change in the status of engineered safety features from automatic to manual mode). If operating procedures are deployed to recover from failures under such FOSs, they are credited in the PSA model. For example, the primary cooling system remains intact during the shutdown state of the reactor. If there is a fault in the decay heat removal system, it is still possible to credit the primary cooling system for resuming core cooling. The primary system may be effective in contributing to the mitigation of the consequences of loss of the decay heat removal function. Before crediting such actions in the PSA model, they have to be properly assessed and included in the facility emergency operating procedures.

The output of the event sequence analysis is a set of accident sequences for core or individual fuel damage and the consequences of experimental facility failures.

3.6.2.7. Human reliability analysis

The human reliability analysis is an important aspect of LPSD PSA Level 1, as the execution of operational and maintenance procedures in support of refuelling, surveillance, testing and repair becomes more complex owing to multiple activities involving multiple agencies. It is good practice to identify human reliability aspects related to various FOSs and quantify them using either facility specific or generic data and performance shaping factors.

3.6.2.8. *Uncertainty, sensitivity and importance analysis*

This step is similar to that in a full power Level 1 PSA (see Section 3.3.10).

3.6.2.9. *Interpretation of results and documentation*

The content of the main report and executive summary is similar to that of the full power Level 1 PSA report, except that the LPSD PSA Level 1 report contains steps on FOS identification, initiating event frequency formulation (including refuelling, test and experiment related initiating events), load drop and system modelling of other systems in the facility, apart from frontline systems, which can be used during low power and shutdown states to meet the safety function requirements (e.g. using a fire hydrant system for alternative cooling).

4. RAM ANALYSIS

This section describes the main applications of RAM analysis with respect to the operation, maintenance, utilization and modification of research reactors.

4.1. INTRODUCTION

In order to utilize a research reactor effectively, the facility needs to be operated safely for the defined mission time(s). Moreover, in order to meet the goals of reliability and availability, the facility needs to have good maintainability characteristics. These three aspects, namely reliability, availability and maintainability, are modelled as part of probabilistic RAM analysis. This section focuses on the specific practices and guidelines relevant to the formulation and implementation of a RAM programme. Differences related to the interpretation of information and data, models and insights, as compared with PSA, are also discussed to bring clarity to the performance of RAM analyses.

The RAM analysis process consists of a number of steps that are similar to PSA, such as system familiarization, management of the process of analysis, identification and selection of initiating events and failures, system modelling, data analysis, quantification of models, formulation of risk metrics (qualitative and/or quantitative), interpretation of results and generation of recommendations.

However, the RAM analysis process varies from the PSA process. As an example, consider a relay based reactor protection system in which, upon receiving a signal from reactor trip logic, the relay is de-energized to open the relay contact in the protection circuit and the shutdown devices are actuated. Failure to open the relay contact in such a case is an unsafe failure and these data are used for PSA modelling. If the relay contact opens for a spurious reason, the failure is safe but affects reactor availability with another statistical datum. Therefore, for RAM analysis, the combined failure data are used. This example illustrates the categorization required for the failure data for PSA and RAM analysis.

4.2. FUNDAMENTALS OF RAM ANALYSIS

There are two important considerations related to the interpretation of component failure:

(1) It is impossible to predict the exact time of failure and, in this context, failures are random in nature.
(2) There can be several modes of failure and this requires analysts to consider which mode of failure forms the definition of failure.

It is often the case that all modes of failure conservatively form the definition of failure. The above random aspect of failure requires that the failure data be treated probabilistically. This is desired as a quantified approach in support of decisions related to engineering management and overcomes the limitations of the qualitative approach in decision making — which is based on engineering and operational judgement.

It is therefore necessary to quantify not only risk, but also reliability, availability and maintainability. The probabilistic methods enable the quantification of these aspects. It also enables the quantification of uncertainty in the estimates, which helps to assess design and safety margins in support of decision making.

4.2.1. Reliability

Reliability (R) is the probability that one component, system or service will perform satisfactorily for a specified period of time under given conditions. This can be expressed mathematically as shown in (4):

$$R(t) = P(T \geq t \,|\, c_1, c_2, c_3 \ldots) \tag{4}$$

where

T is the random variable representing actual time to failure of the component;
t is the mission time (for non-repairable components);

and $c_1, c_2, c_3 \ldots$ represent the operational and environmental conditions, such as temperature, humidity, electrical stresses and lubrication.

Considering the case of constant failure rate, in which an exponential distribution can be used for modelling, the reliability is estimated as shown in (5):

$$R(t) = \exp(-\lambda \times t) \tag{5}$$

where

λ is the failure rate (per unit time);

and t is the mission time (unit time).

For the case of exponential distribution in which the failure rate is considered as constant, the inverse of failure rate is the mean time to failure (MTTF) and is expressed in a unit of time, such as hours (6).

$$\text{MTTF} = \frac{1}{\lambda} \tag{6}$$

For repairable systems, the parameter mean time between failures (MTBF) is used, as it is assumed that after repair the component will be available to perform its function until it fails again. The life cycle of the component is then characterized by uptime and downtime.

This subsection presented some introductory aspects of reliability, in which exponential distribution was assumed as the applicable distribution. However, reliability estimation may require analysis of the data, selection of a suitable probability distribution and assessment of the failure rate.

4.2.2. Availability

Availability is defined as the fraction of time during which the component is capable of fulfilling its intended purpose. Availability, when it is associated with design aspects, is referred to as an inherent availability, while in the operational domain it is referred to as the steady state or transient availability. In terms of uptime and downtime parameters, the operational availability is defined as follows (7):

$$\text{Availability} = \frac{\text{uptime}}{\text{uptime} + \text{downtime}} \tag{7}$$

Steady state availability, in terms of MTBF and mean time to repair (MTTR), can be expressed as shown in (8):

$$\text{Availability (steady state)} = \frac{\text{MTBF}}{\text{MTBF} + \text{MTTR}} \tag{8}$$

The availability of SSCs can be improved through design (e.g. provision to facilitate automatic recovery by the identification and isolation of faulty components and switching over to healthy redundant components). Similarly, effective surveillance and periodic testing of SSCs reveal latent faults and the corrective actions taken will improve availability.

4.2.3. Maintainability

Maintainability is the probability that a failed service or system or component will be repaired or restored to service in a given period of time when maintenance is performed by trained and qualified persons in accordance with the prescribed procedure. The expression for maintainability, assuming an exponential distribution of the time to repair, may be given as follows (9):

$$M(t) = \exp(-\mu \times t) \tag{9}$$

where

μ is the repair rate (i.e. the number of repairs per unit of time);

and t is the repair time.

MTTR is also defined as shown in (10) and is also referred to as recovery time.

$$\text{MTTR} = \frac{1}{\mu} \tag{10}$$

The design has to ensure that provisions of maintainability and inspectability are covered, which includes, inter alia, ease of inspection, repair and replacement, adequate access to the components, adequate provision for the detection and isolation of failed components, shielding from radiation and spare parts inventory management. Automation such as self-diagnostic features also improves maintainability as it minimizes the human factor.

The designer of the systems therefore has to decide on the level of automation, particularly in control systems, to realize higher maintainability. This requires careful review of the reduction of dependence on human factors, but on the other hand calls for careful assessment of safety and availability issues. The safety channels have to be designed in such a way that all computer systems and communication links are poised for operation and available to perform the postulated function on demand. This requires automatic reconfigurable features, along with periodic testing facilities and enablers. The automatic reconfigurable feature will recover any failure during the mission, while the testing and condition monitoring features will ensure higher maintainability and availability. The major objective when designing the control system is that, during an emergency situation, the cognitive load on the operator is as low as reasonably achievable, whereby information related to malfunctions or failures is made available through the human–machine interface in a manner that enables the operator to recover the situation effectively.

An effective maintenance programme is a major part of ensuring higher reliability and availability of components during the operational stage of the reactor. Proactive maintenance, which includes schedule based preventive maintenance or condition based predictive maintenance, forms an integral part of maintenance management.

A risk based approach for maintenance, periodic testing and inspection also provide a quantitative method for prioritizing and optimizing these activities (e.g. test or service intervals).

4.3. PROJECT FRAMEWORK FOR THE APPLICATION OF RAM ANALYSIS

Figure 1 shows a project framework for the application of RAM analysis to research reactors. The procedure starts with the definition of the objective and scope of the project, and continues with facility and data information collection, identification of contributors to system unavailability, system modelling, identification and prioritization of SSCs, comprehension of results, recommendations related to health management of SSCs, and feedback for tuning the project.

In PSA, the frequency of PIEs is important for safety assessments, while in RAM analysis the contribution of the PIEs to unavailability is investigated. Therefore, to initiate RAM analysis, the list of PIEs available from the PSA needs to be revised to assess the potential contribution of the PIEs to facility unavailability. An example is an event of off-site (class IV) power failure, which forms part of the PIEs in PSA and, at the same time, affects facility availability. Such events also form part of RAM analysis, and here the concern is how to reduce facility downtime. Moreover, external events tend to have greater consequences in terms of safety and availability. Therefore, a complete review is required to assess those aspects that contribute to system unavailability, such as refuelling, maintenance schedules, component or system failures, surveillance and test programmes. Some failures of equipment or process systems may not have safety consequences but may be important from the point of view of availability. For example, failure of the fuelling equipment or machine to start refuelling the core positions may affect facility availability. However, failure to ensure the cooling of fuel during fuel handling has safety consequences.

Similarly, safety system failure, whether safe or unsafe, also affects facility availability. This aspect is analysed as part of RAM analysis. Special attention is paid to the human error aspects that cause reactor shutdown or failure in the support systems that finally leads to facility or system unavailability. Here, facility records, particularly incident reports and maintenance logs, are reviewed to assess the potential for human induced failures that contribute to reactor trips or unavailability.

The application of RAM analysis to research reactors requires the adoption or modification of the models available for nuclear power plants. In principle, RAM analysis can be performed at the design stage, and more appropriately during the conceptual design stages, at the operational stage or even in support of an ageing management programme.

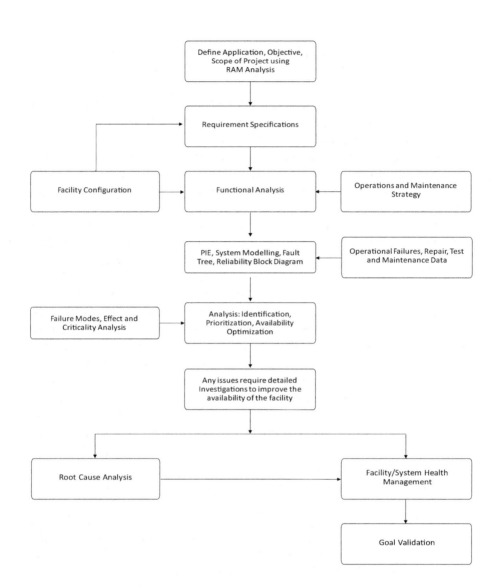

FIG. 1. Project framework for the application of RAM analysis.

4.4. MAJOR ELEMENTS OF RAM ANALYSIS

4.4.1. Objective and scope

The objective of a project using RAM analysis, such as whether the project deals with design optimization and evaluation, operating facility optimization of availability, or enhancement of maintainability, have to be clearly defined. The objective may vary depending on whether RAM analysis is being performed at the conceptual design stage, design stage, operational stage or in support of ageing management. When RAM analysis is performed at the conceptual or design stage, the objective is an overall availability target (e.g. a target availability of 80%). In such situations, the analysis is performed using generic reliability data (e.g. failure rate, MTTR, test intervals).

For the purpose of this publication, it is assumed that RAM analysis is performed at the operational stage of the facility. At this stage, the main objective of the activity is to improve facility availability and the secondary objective is to improve system availability by identifying the major contributors to system unavailability.

The results of RAM analysis can be used for an availability improvement strategy (e.g. the implementation of a reliability centred maintenance programme), and thereby focus on those areas that contribute significantly to availability improvement. The results can also be used in support of identification and prioritization of maintenance and surveillance activities.

The following aspects determine the scope of RAM analysis:

(a) Level. Whether the analysis is performed at facility or system level, or it only deals with a small subsystem. If RAM analysis is performed at facility level, the analysis of initiating events, along with systems analysis, forms part of the RAM analysis. However, if RAM analysis is performed at system level, the analysis of initiating events may not form part of the RAM analysis.
(b) Depth. Whether the analysis needs to be qualitative or quantitative.
(c) System boundary. This needs to be defined clearly, together with the input and output relationship with interfacing systems in terms of shared hardware, control of information flow or administrative controls.
(d) Special issues. These include the treatment of uncertainties and major assumptions, human error modelling, common cause failure modelling and sensitivity analysis, and termination point, such as the presentation of the recommendations or the implementation of identified improvements and feedback programmes.

4.4.2. Project management for RAM analysis

The major management features for a RAM analysis project are as follows:

(a) The team involved in RAM analysis comprises a team leader, who is conversant with the development of the application of RAM analysis for research reactors, and other team members who:
    (i) Have domain knowledge — an intimate knowledge of the system design and operations;
    (ii) Have the necessary qualifications and training in reliability engineering;
    (iii) Are fully conversant with the facility maintenance and surveillance programmes for mechanical, electrical, and instrumentation and control systems.
(b) The line of authority and communication has to be clearly defined for smooth execution of the project. The team leader may report to the facility manager. Communication with the maintenance agencies needs to be continuous on aspects related to data collection, system modelling and interpretation of the results of the analysis.
(c) The quality assurance plans and procedures have to be developed for the various steps of the RAM analysis project.
(d) Reviews have to be performed and documented at various levels and stages. A first level review is performed by the team leader to check that the analysis is numerically correct and in line with the scope and objective of the project. A second level review is performed by the design and operations teams. Comments are recorded in a manner that facilitates external or peer review. The objective of these third level reviews is to ensure that the analysis is numerically correct and complete to the fullest extent possible and addresses the issues in an effective manner.
(e) Documentation forms an important aspect of the analysis. The objective is that all of the technical details are traceable to the basic level.

The success of the RAM analysis project depends to a large extent on the availability and adequacy of resources, including the following:

(a) Number of trained staff in the areas of design, operation, reliability and safety engineering, maintenance management and root cause analysis;
(b) Infrastructure and administrative support;
(c) Software tools, such as for reliability modelling, data analysis and managing various levels of reviews;
(d) Project funding that can meet capital costs as well as administrative costs.

The resources have to include not only the cost of RAM analysis development and implementation, but also resources required for running and maintaining the project.

4.4.3. Facility familiarization

Each member of the project team has to clearly understand the objective and scope of the RAM analysis. The system familiarization programme aims to introduce or update the team with respect to all design, operational and safety aspects of the facility. The familiarization programme for various individual systems has to include, as necessary, formal classroom lectures on technical documentation, including design basis reports, technical specifications, procedures and schedules for radiation protection, operation and maintenance, facility operational limits and conditions, and operational manuals. The depth of subject knowledge is commensurate with RAM analysis modelling needs. Training on subjects such as data collection and analysis, reliability software, and requirements related to documentation and administration also forms part of the familiarization programme.

4.4.4. Functional analysis

Functional analysis is carried out to systematically understand the functional relationships among the systems and subsystems, such that an integrated RAM analysis model can be created for the complete facility. A functional analysis approach is very effective for systems being designed with new concepts and new technology. Functional analysis provides an overview of various facility functions, including support functions to meet the facility's overall function, subfunctions to support the main functions, and so on, until the component level and level of associated supporting human actions are reached. The level of detail is a function of the objective and scope of the analysis. System requirements form the basis of functional analysis, while the outcome of functional analysis includes system design synthesis, modes of system operations, limits and conditions. There are many tools and methods for functional analysis.

4.4.5. Models for RAM analysis

The procedure for identifying the initiating events and various SSC failure modes in RAM analysis is similar to defining events for a PSA, except that for RAM analysis these events are seen in the context of facility and system availability. The approach used to formulate the list of initiating events can be found in Section 3.3.4.

There are three major approaches for RAM modelling. These include failure mode, effects and criticality analysis (FMECA), reliability block diagrams and fault tree analysis. There are many other approaches, such as Markov chain models, which are employed for dynamic modelling of systems, but the scope of this subsection is limited to the discussion of static approaches.

4.4.5.1. *Failure mode, effects and criticality analysis*

This is a qualitative approach to systems analysis. The FMECA process involves systematically identifying each component of the reactor system one by one and recording, in a tabular format, various modes of the component failure, its effects at local, system and facility levels, assignment to this mode of failure of a qualitative ranking for likelihood and, finally, assignment of a criticality ranking. Furthermore, determining whether a provision exists to identify and isolate this failure and whether there is a mechanism for recovery also forms part of FMECA. This analysis is more suited to qualitative evaluation of component failure modes, their effects at local and global or facility levels, the recovery mechanism to compensate for the failures and, finally, prioritization of the failure modes with respect to contribution to facility unavailability.

The main procedure for performing FMECA as part of the RAM analysis is similar to that followed in PSA. However, the objective of RAM analysis is to assess the consequences in terms of loss of facility or system availability, whereas the outcome of FMECA is the identification of components and subsystems that are important for facility availability. Even though the FMECA process is qualitative in nature, the classification of components using a categorization based on qualitative indicators, such as the frequency of occurrence of events (e.g. negligible, very low, low, medium, high, very high), along with the severity of consequences categorized on similar lines, enables not only identification but also prioritization of components and systems that are important for availability.

The output of FMECA is used to develop a qualitative criticality matrix, as shown in Fig. 2. This matrix is used to develop maintenance strategies that allow components that fall in an unacceptable zone to be brought into an acceptable zone. This matrix provides a graphical representation of the performance of a component in terms of strengthening the maintenance programme. At the same time, it also provides a rationale for tracking the performance of components in the acceptable zone.

4.4.5.2. *Reliability block diagram*

Reliability block diagrams are one of the main approaches used for RAM analysis modelling. The limitation of the FMECA approach is that it is

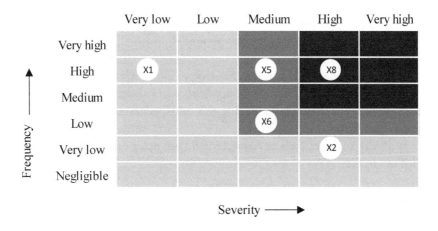

FIG. 2. Qualitative criticality matrix. (X1, X2 ... represent failure mode examples.)

not capable of capturing interactions between two or more components, as it considers system reliability or availability in terms of the selection and analysis of a single component failure. A reliability block diagram is similar to a fault tree and captures complex component and subsystem dependencies for system level modelling. Key outputs of reliability block diagrams include the identification of system and component boundaries, the input–output relationship at component and system levels, and functional dependencies, as well as the design for system configuration (particularly the redundancy levels). The major advantage of this approach is that the arrangement of the blocks often represents the system configuration and it is therefore easy to comprehend the functioning and dependencies of the system.

4.4.5.3. Fault tree analysis

Fault tree analysis for RAM analysis is similar to that applied for system modelling in Level 1 PSA, as presented in Section 3.3.7.

4.4.6. Data assessment and parameter estimation for RAM analysis

The data requirements for RAM analysis are slightly different from those for PSA. For example, in PSA, one of the key parameters that determine system unavailability is probability of failure on demand. This parameter is considered to be a time independent failure. Demand failure probability for a safety system refers to failures that remain passive until demand is put on the system to perform, and it is a function of the standby failure rate of the system and the test or demand interval. In PSA, assessment of the failure probability of a component is carried out using a time averaged standby failure model, in which standby failure λ_s and test interval t form the input for assessing the failure probability or unavailability of standby components. However, these types of data and parameters are not directly applicable to RAM analysis applications. This example has been given here to alert RAM analysts to distinguish and understand the data for RAM analysis applications.

In RAM analysis, data assessment and parameter estimation are generally carried out in four major steps:

(1) Event definition;
(2) Model parameter selection;
(3) Identification of data source;
(4) Parameter estimation.

These steps are applicable to the assessment of failure frequency for initiating events, as well as failure probabilities for components.

The definition of an event has to include the applicable failure mode, failure criteria, component boundary and associated assumptions, if any. The component has to be described by the mode of its operation, such as on-line repairable or non-repairable, standby tested and monitored. Often it is useful to identify event groupings. This is particularly useful for the initiating events.

The parameters vary according to the type of component or event and need to be identified. Table 7 shows the parameters and models for various categories of components used in RAM analysis.

It is assumed that the facility will not have any components that cannot be tested or monitored, hence a model for non-tested standby components has not been given. If a component cannot be tested, an engineering modification or in-service surveillance programme has to be drawn up to ensure that the condition of the component can either be assessed periodically or that suitable parameters that provide the status of the components are monitored on-line or at suitable intervals.

For each model parameter, associated data requirements have been indicated in Table 7. To derive these parameters in the model, it has been assumed that the data follow an exponential distribution. This means that the failure rates are constant. In cases in which the determination of a different distribution for a given data set is required, and the failure rate shows an increasing or decreasing trend, the Weibull distribution or any other appropriate distribution can be used.

TABLE 7. UNAVAILABILITY MODELS, PARAMETERS AND DATA REQUIREMENTS FOR VARIOUS CATEGORIES OF COMPONENTS

Category of event and component and associated unavailability mode	Unavailability model	Parameter	Data requirement
On-line components			
Non-repairable	$U_{n-r} = 1 - \exp(-\lambda_o t_m)$	λ_o — operating failure rate (failures/hour); t_m — mission time, time for which the component is supposed to operate to meet the success criteria or mission	$\lambda_o = N_o / T$ N_o — number of failures in operating condition in a given time T Observed mission time to obtain t_m
Repairable	$U_r = \dfrac{\lambda_o t_r}{1 + \lambda_o t_r}$	λ_o — operating failure rate (failures/hour); t_r — mean time to repair in hours	$\lambda_o = N_o / T$ N_o — number of failures in operating condition in a given time T Observed time to repair to obtain t_r

TABLE 7. UNAVAILABILITY MODELS, PARAMETERS AND DATA REQUIREMENTS FOR VARIOUS
CATEGORIES OF COMPONENTS (cont.)

Category of event and component and associated unavailability mode	Unavailability model	Parameter	Data requirement
Tested standby			
Hardware failure	$U_{st\text{-}h} = 1 - \dfrac{1 - \exp(-\lambda_s t_i)}{\lambda_s t_i}$	λ_s — standby failure rate (failures/hour); t_i — test interval in hours	$\lambda_s = N_s / T$ N_s — number of failures in standby condition in a given time T (revealed when there is a need for testing a system) Observed test interval to obtain t_i
Test outage	$U_{t\text{-out}} = \dfrac{t_d}{t_i}$	t_d — test duration in hours; t_i — test interval in hours	Observed test duration to obtain t_d Observed test interval to obtain t_i
Repair outage	$U_{r\text{-out}} = \lambda_s t_r$	λ_s — standby failure rate (failures/hour); t_r — mean time to repair in hours	$\lambda_s = N_s / T$ N_s — number of failures in standby condition in a given time T (revealed when there is a need for testing a system) Observed time to repair to obtain t_r

TABLE 7. UNAVAILABILITY MODELS, PARAMETERS AND DATA REQUIREMENTS FOR VARIOUS CATEGORIES OF COMPONENTS (cont.)

Category of event and component and associated unavailability mode	Unavailability model	Parameter	Data requirement
Scheduled maintenance	$U_m = t_{mt} f_m$	t_{mt} — mean time of maintenance in hours; f_m — frequency of maintenance/ hour	Observed time to maintenance to obtain t_{mt} f_m obtained from facility record
Monitored standby components	$U_{st-mon} = \dfrac{\lambda_s t_r}{1 + \lambda_s t_r}$	λ_s — standby failure rate (failures/hour); t_r — mean time to repair in hours	$\lambda_s = N_s / T$ N_s — number of failures in standby condition in a given time T (revealed when there is a need for testing a system) Observed time to repair to obtain t_r

The final column of Table 7 shows the data requirements for parameter estimation. It is assumed that a reliable and structured data collection system is in place. The major features of this system are the following:

(a)   A component coding system that facilitates component categorization (e.g. based on the type of component, the reactor system it belongs to, applicable failure modes, safety/non-safety), which will help in effective categorization and utilization of the database.
(b)   Categorization of data as safety related or availability related.
(c)   Definition of failure. For example, the definition of failure for a standby diesel generator could be failure to start on demand or failure to run for the designated mission time. Both of these failures are related to safety. The unavailability contribution owing to scheduled preventive maintenance or unscheduled repair is an availability concern.
(d)   The amount of data coming from testing and surveillance. For example, if the diesel generators are tested, they have to be loaded to the specified minimum loads to constitute a valid demand for assessing the number of demands on diesel generators.
(e)   Data collection by staff dealing in specific domains. For example, instrument maintenance staff are responsible for collecting data on instrument component or system failure.
(f)   Guidelines on categorization of non-repairable and repairable components.
(g)   Development of general guidelines on how to detect failure, what constitutes a failure, partial failure and how to evaluate the repair or recovery time and address aspects related to detection time. For example, a relay was detected as having failed during one of the demands on the system during a night shift. The maintenance staff replaced the relay in the morning after obtaining the work permit. Clear guidelines as to which period is to be considered for allocation as repair time are required. Was it from the time that the relay failed during the night shift until it was replaced or from the time that the permit was obtained until the relay was replaced? The reliability engineer has to develop clear guidelines on the selection of different test and maintenance models.

The software to be used for RAM analysis has to be verified. The software has to offer models and methods for the following:

(a)   Compilation of component reliability database;
(b)   FMECA;
(c)   Reliability block diagrams;
(d)   Importance and sensitivity analysis;

(e) Uncertainty evaluation;
(f) Common cause failure modelling;
(g) Markov chain analysis.

4.4.7. Uncertainty, sensitivity and importance analysis

The procedure for uncertainty, sensitivity and importance analysis is similar to that described in Section 3.3.10 for Level 1 PSA.

Importance analysis is carried out in support of the prioritization of components. Many models are available for importance ranking. Which model is selected depends on the application. The basis for the selection of models for a given application has to be documented.

4.4.8. Requirements of RAM analysis

The major objective of this stage is to ensure that the facility model meets the defined reliability and availability requirements of the RAM analysis. This can be achieved by optimizing system configurations through the use of redundancy, optimization of maintainability and inspectability. For example, the system or facility level fault tree model for unavailability is used as a tool to allocate RAM analysis parameters such that desired goals are achieved. During the design stage, this process helps to achieve inherent facility availability by the following:

(a) Employing strategies for minimizing repair time, enhancing inspectability, incorporating provision of automatic testing and using components that have higher reliability;
(b) Optimizing configuration (e.g. by incorporating redundancy and diversity);
(c) Improving logistics and inventory management;
(d) Minimizing the human factors in testing and maintenance.

This is an iterative process that starts from allocations for top event availability and moves down to component level availability.

Recovery plays a major role in availability allocations. A small change in recovery times can change the top event availability significantly. Another factor that is important is coverage probabilities, which are related to the ability of a system to detect and locate a fault. Detection, location, inspection and diagnosis coverage affect the efficiency of the RAM analysis and availability goals.

The availability allocation process generates input for surveillance, testing and maintenance schedules. Traditionally, the outage schedules in research reactors are generated based on operational experience and engineering judgements. The RAM analysis project optimizes the testing, maintenance and overall surveillance programme using a systematic approach.

4.5. ROOT CAUSE ANALYSIS

In many cases, detailed root cause analyses of various failure modes are required to understand associated failure mechanisms, including human factors that cause component failure. This is particularly the case for recurring failures that contribute significantly to system unavailability. Root cause analysis is a resource consuming activity, and therefore it has to be limited to select cases in which the gains or benefits justify the resources.

The RAM analysis is recursive in nature. The feedback obtained from the analysis forms an input for improving the RAM specifications and thereby facility or system availability.

4.6. RAM ANALYSIS RESULTS AND RECOMMENDATIONS

Taking into account the scope of the RAM analysis, a certain level of detail (e.g. component, system or facility level analysis) may be included and the RAM specifications can be determined. However, the major specifications that are common in most RAM analyses are as follows:

(a) Statement of facility or system reliability.
(b) Inherent availability of the facility or system.
(c) Overall availability of the facility or system: fraction of time that the reactor is available to produce neutron flux over the scheduled operation time.
(d) Target utilization factor for the following:
   (i) Test loops;
   (ii) Isotope production facilities;
   (iii) Beam experiments.
(e) Statement of facility or system maintainability.
(f) Optimal maintenance interval for system and components.
(g) Type of maintenance.
(h) Optimal surveillance or inspection intervals for system and components.
(i) Statement of optimal dose consumption.
(j) Statement of total maintenance cost of the following:
   (i) Personnel;
   (ii) Machine;
   (iii) Material.

These specifications can further be utilized for the following:

(a) Drafting the operational schedule of the facility;
(b) Shutdown scheduling;
(c) Refuelling planning;
(d) Availability allocation for components and systems.

If the above mentioned statements are generated during the design stage of the facility, these metrics become target specifications. If these statements are generated during the operational stage of the facility, they can be compared with the design stage goals and provide areas for improvements of RAM specifications.

4.6.1. Discussion, interpretation and presentation of results

Similar to PSA, RAM analysis provides the following results:

(a) The identification and prioritization of components or systems or human actions that are important for facility or system availability;
(b) The risk metrics that indicate which activity falls in which criticality zone.

A risk matrix depicts an item's criticality based on the frequency and severity index (or parameters). Discussion includes guidelines on how to use these results in areas such as shutdown scheduling, maintenance optimization and assessment of surveillance frequencies.

4.6.2. Documentation

This step is similar to that described for Level 1 PSA in Section 3.3.11. The table of contents for the Level 1 PSA report provided in Appendix I can be adapted to reflect the typical table of contents for the RAM analysis report.

5. TRAINING AND EDUCATION ON THE USE OF PROBABILISTIC METHODS

Training and education on the application of probabilistic methods for research reactors could have different approaches depending on the target audience (e.g. regulatory body, operating organization or other institution). Since some professionals may be familiar with the design and operational features of research reactors but not with probabilistic methods, while others may be familiar with the methods but not with the facilities, emphasis needs to be placed on different areas of knowledge accordingly. Training on the design features and operational aspects, as well as the operating limits and conditions, of research reactors is to be provided to any group of professionals. Furthermore, taking into consideration the different types of research reactors and their associated utilization, the application of a graded approach to the use of probabilistic methods could also be presented in a training programme.

Whatever the technical field of the professional may be, trainees could benefit from studying examples of past probabilistic method applications for research reactors. Previous operational experience at nuclear facilities needs to be reviewed and incorporated into a training programme, as appropriate. In addition, to gain a better understanding of the objectives and scope of a project using probabilistic methods and the means to fulfil them, it is useful to discuss the main criticisms of similar analyses and how they were responded to.

Additional information and training may be required in the following areas, which are specific to research reactors:

(a) Reactivity and criticality management;
(b) Core thermal conditions;
(c) Safety of experimental devices;
(d) Modification of reactors;
(e) Manipulation of components and materials;
(f) Safety measures for visitors.

5.1. TRAINING OF OPERATING PERSONNEL AND REGULATORY STAFF

Organizations responsible for operating research reactors need to have established training and retraining programmes for their operating personnel. The operating personnel need to be aware of the specifics related to their own facility and to research reactors in general from the point of view of safety and operational aspects.

In such cases, topics in the training programmes related to the probabilistic methods for research reactors familiarize the operating personnel with the application of probabilistic methods for safety assessment and for the reliable operation of their facilities. When a probabilistic method is developed in the FOS, it is possible to involve reactor operating personnel from the early stages of the development. Moreover, applications of probabilistic methods rely on a robust and user friendly database, which in turn facilitates the subsequent use of the probabilistic method for the research reactor as a complementary tool. The use of a computer program for simulating accident sequences enhances familiarity with probabilistic method modelling.

Maintenance staff can participate in the development of the probabilistic method by logging facility performance data specifically identified for these studies. Such failure and operational performance data are noted in a format suited to probabilistic methods and include data such as dependent failures (common cause/common mode) and human factors as particular entries in the log. This involvement may foster an interest in safety, in such a way that reactor operators remain confident that their research reactor still conforms to the original licensing requirements.

A risk monitor is designed to be used by all facility personnel, rather than just by specialists of probabilistic methods, and the user does not need to have special knowledge of or training in probabilistic methods. The changes that the user can make to the facility configuration are limited, for example, to specifying the FOS and identifying the components that have been removed from service for maintenance purposes. This is done using the normal facility identifiers for the equipment selected. If, on the other hand, the user is trained to interact directly with the probabilistic model of a risk monitor, the benefits of the application of this tool increase substantially.

Safety culture training is an opportunity for the participants of a facility focus group to increase their understanding of probabilistic concepts. The integration of PSA techniques into safety culture training could facilitate the achievement of the main purposes of these activities, which include looking at potential responses to safety problems, highlighting the need for improvements to be made in the management of safety and identifying the most appropriate solutions.

Regulatory staff need to be provided with training to improve their understanding of the benefits of applying probabilistic methods to regulatory issues as a complementary tool for deterministic analyses. Furthermore, training and education in probabilistic methods would serve to highlight operational aspects of research reactors warranting particular attention during the licensing process, as well as throughout the entire process of regulatory supervision of the facility.

Training on the interpretation of the numerical significance of the probabilistic data and results and their importance to the everyday operation of the facility and to accident conditions is also to be undertaken.

5.2. TRAINING OF PROBABILISTIC METHOD PRACTITIONERS

A group of analysts applying a probabilistic method to a research reactor (PSA and/or RAM analysis) for the first time will require training to acquire the expertise necessary to complete the project successfully. This expertise can vary in depth, depending on the scope of such a project and the participation of the facility designer and the operating personnel. Even if the practitioner already has some of the expertise required, training for familiarization with the facility features, as well as on the objectives, procedures and methods of the project, is beneficial. The training can be organized according to the required expertise, whether facility related or probabilistic method related.

As a minimum, the following three training modules are to be attended by all practitioners:

(1) Module 1 — facility systems and operating procedures. This module has to cover the basic aspects of system design, including operating procedures under normal and accident conditions. A condensed version of the corresponding training courses given to reactor operating personnel and engineering personnel is an example of such a module. The objectives of this module are to give practitioners an understanding of facility behaviour and to emphasize the complexity of research reactors with respect to core management, human factors, experimental programmes and utilization.

(2) Module 2 — probabilistic methods. This module has to elaborate on the relationship between PSA and RAM analyses, covering issues such as event sequence and system modelling (e.g. event trees and fault trees), quantification of accident sequences, availability analysis, uncertainty, sensitivity and importance analysis, data handling and human reliability analysis. The objective of this module is to introduce the practitioners to the special methodological problems and techniques involved, together with specific issues concerning the software to be used in the analysis. The focus of the module is on establishing a common understanding of the techniques in order to resolve possible misconceptions about the advantages and disadvantages of the various methods. The module is ideally attended by appropriate subgroups (PSA and RAM analysis) and addresses the graded approach to the use of probabilistic methods for research reactors, based on their potential hazard.

(3) Module 3 — probabilistic method procedures and comparative reviews of the topic using similar facilities. This is an important part of the preparation of a group of probabilistic method practitioners. The module consists of a review of the complete assessment procedure and comparative reviews of PSA and RAM analysis that have been performed for facilities of a similar

design. The objective of this module is to enhance a common understanding of PSA and RAM analysis among the practitioners.

Complementary to these three modules, PSA and RAM analysis training can be consolidated by conducting a pilot study that consists of the development of accident sequences and associated fault trees for a specific initiating event. This activity exposes the practitioners, in a short period of time, to most of the issues to be faced when they are conducting the entire project.

The recommended contents of a training programme for probabilistic methods are presented in Table 8. The level of detail and the intensity of training need to be adapted to the requirements of the target group (e.g. probabilistic method practitioners, operating personnel, engineers, regulators, vendors).

TABLE 8. TRAINING PROGRAMME FOR PROBABILISTIC METHODS

Main topics	Main subtopics
Reliability theory and applications (PSA and RAM analysis)	• Overview of reliability concepts/dependability and related concepts • System reliability, availability and maintainability • Design for reliability • FMEA/FMECA • Reliability block diagrams • Reliability data estimation/reliability testing/component qualification

PSA for research reactors	• Risk/safety issues for research reactors
	• Graded approach to safety analysis for research reactors
	• Level 1 PSA
	  — Initiating event identification
	  — Event tree/fault tree analyses
	  — Common cause failures for Level 1 PSA
	  — Human reliability analysis for Level 1 PSA
	  — Level 1 PSA quantification issues
	  — Risk importance measures
	  — Results of Level 1 PSA: core damage frequency
	  — Level 1 PSA: practical issues, learning and insights

TABLE 8. TRAINING PROGRAMME FOR PROBABILISTIC METHODS (cont.)

Main topics	Main subtopics
	• Level 2 PSA — Severe accident phenomena — Deterministic safety analysis and Level 2 PSA — Level 1/Level 2 PSA interface (core damage states) — Accident progression event trees for Level 2 PSA — Level 2 PSA quantification issues — Results of Level 2 PSA: probabilities of release categories and source terms — Severe accident management
PSA for research reactors (cont.)	• Level 3 PSA — Level 2/Level 3 PSA interface — Results of Level 3 PSA: risk estimates — Emergency response planning (off-site accident management and economic impacts)
Uncertainties in PSA and sensitivity studies	• Issues and types of uncertainty in PSA • Sources of uncertainty in assessment • Uncertain data and uncertainty propagation • Application of Bayesian methods • Sensitivity analysis • Importance analysis
Integrated risk informed decision making and application of PSA	• Concepts and implementation • Risk informed inspection • Level 2 PSA and facility improvement • PSA and decision making • Probabilistic safety criteria • Application of Level 2 PSA results to source term prediction

Note: FMEA — failure mode and effect analysis.

Appendix I

TYPICAL TABLE OF CONTENTS FOR A LEVEL 1 PROBABILISTIC SAFETY ASSESSMENT REPORT

EXECUTIVE SUMMARY

MAIN REPORT

APPENDICES TO THE MAIN REPORT

C.   Human reliability analysis

D.   Dependence analysis

E.   Quantification of accident sequences

Appendix II

EXAMPLES OF FACILITY OPERATIONAL STATES
FOR HIGH POWER RESEARCH REACTORS

TABLE 9. EXAMPLES OF FACILITY OPERATIONAL STATES FOR
HIGH POWER RESEARCH REACTORS

FOS	FOS-No.	State	Characteristic
Power	FOS-0	Steady state	Reactor power — 1% of full power to 100%; coolant system pressure P_o >6–12 kg/cm^2; coolant temperature T_o <50°C; T_{max} 70°C; class IV power supply normal; all shutdown rods up; heat removal — primary coolant system
Power lowering	FOS-1	Transition	Reactor power being lowered in auto/manual mode; reactor power — >1% of full power; coolant system pressure P_o >6–12 kg/cm^2; coolant temperature T_o <70°C; all shutdown rods up; heat removal — primary coolant system
Low power	FOS-2	Steady state	Reactor at low power (~1% of full power); coolant system pressure P_o >6–12 kg/cm^2; coolant temperature T_o <40°C; all shutdown rods up; heat removal — primary coolant system
Shutting down	FOS-3	Transition	Reactor is being taken to shutdown state; reactor at low power (<1% of full power); coolant system pressure P_o >6–12 kg/cm^2; coolant temperature T_o <40°C; all shutdown rods up; heat removal — primary coolant system
Shutdown with primary pump on	FOS-4	Steady state	Reactor is in normal shutdown state; reactor power in kW — tens of watt range; coolant system pressure P_o >6–12 kg/cm^2; coolant temperature T_o <40°C; heat removal — primary coolant system
Shutdown with primary pump off	FOS-5	Steady state	Reactor shutdown with shutdown of main coolant pump — reactor is in normal shutdown state; reactor power in kW — tens of watt range; coolant system pressure P_o >0.2–3 kg/cm^2; coolant temperature T_o <40°C; heat removal — shutdown cooling system or decay heat removal system

TABLE 9. EXAMPLES OF FACILITY OPERATIONAL STATES FOR HIGH POWER RESEARCH REACTORS (cont.)

FOS	FOS-No.	State	Characteristic
Shutdown fuelling	FOS-6	Steady state	Reactor shutdown with primary pump off; reactor is in normal shutdown state; reactor at low power, in kW — tens of watt range; coolant system pressure P_o >0.2–3 kg/cm^2; coolant temperature T_o <40°C; heat removal — shutdown cooling system or decay heat removal system
Startup	FOS-7	Transition	Reactor startup to reach low power operation; reactor power <1% of full power; reading of log rate ~1–2%; reactor control startup manual or auto and linear power on scale; coolant temperature T_o >50°C; T_{max} 70°C; class IV power supply normal; all shutdown rods up; heat removal — primary coolant system
Low power	FOS-8	Steady state	Reactor at low power (~1% of full power); coolant system pressure P_o >6–12 kg/cm^2; coolant temperature T_o <40°C; all shutdown rods up; heat removal — primary coolant system
Power raising	FOS-9	Transition	Reactor power above 1% of full power to 100%; coolant system pressure P_o >6–12 kg/cm^2; coolant temperature T_o >50°C; T_{max} 70°C; class IV power supply normal; all shutdown rods up; heat removal — primary coolant system

REFERENCES

[1]   INTERNATIONAL ATOMIC ENERGY AGENCY, Safety of Research Reactors, IAEA Safety Standards Series No. SSR-3, IAEA, Vienna (2016).

[2]   INTERNATIONAL ATOMIC ENERGY AGENCY, Safety Assessment for Facilities and Activities, IAEA Safety Standards Series No. GSR Part 4 (Rev. 1), IAEA, Vienna (2016).

[3]   INTERNATIONAL ATOMIC ENERGY AGENCY, Optimization of Research Reactor Availability and Reliability: Recommended Practices, IAEA Nuclear Energy Series No. NP-T-5.4, IAEA, Vienna (2008).

[4]   INTERNATIONAL NUCLEAR SAFETY ADVISORY GROUP, Basic Safety Principles for Nuclear Power Plants 75-INSAG-3 Rev. 1, INSAG Series No. 12, IAEA, Vienna (1999).

[5]   INTERNATIONAL ATOMIC ENERGY AGENCY, Probabilistic Safety Assessment for Research Reactors, IAEA-TECDOC-400, IAEA, Vienna (1987).

[6]   INTERNATIONAL ATOMIC ENERGY AGENCY, Application of Probabilistic Safety Assessment to Research Reactors, IAEA-TECDOC-517, IAEA, Vienna (1989).

[7]   INTERNATIONAL ATOMIC ENERGY AGENCY, Manual on Reliability Data Collection for Research Reactor PSAs, IAEA-TECDOC-636, IAEA, Vienna (1992).

[8]   INTERNATIONAL ATOMIC ENERGY AGENCY, Reliability Data for Research Reactor Probabilistic Safety Assessment. Final Results of a Coordinated Research Project, IAEA-TECDOC-1922, IAEA, Vienna (2020).

[9]   INTERNATIONAL ATOMIC ENERGY AGENCY, Development and Application of Level 1 Probabilistic Safety Assessment for Nuclear Power Plants, IAEA Safety Standards Series No. SSG-3, IAEA, Vienna (2010).

[10]   INTERNATIONAL ATOMIC ENERGY AGENCY, Development and Application of Level 2 Probabilistic Safety Assessment for Nuclear Power Plants, IAEA Safety Standards Series No. SSG-4, IAEA, Vienna (2010).

[11]   INTERNATIONAL ATOMIC ENERGY AGENCY, Regulatory Review of Probabilistic Safety Assessment (PSA) – Level 1, IAEA-TECDOC-1135, IAEA, Vienna (2000).

[12]   INTERNATIONAL ATOMIC ENERGY AGENCY, Regulatory Review of Probabilistic Safety Assessment (PSA) – Level 2, IAEA-TECDOC-1229, IAEA, Vienna (2001).

[13]   INTERNATIONAL ATOMIC ENERGY AGENCY, Determining the Quality of Probabilistic Safety Assessment (PSA) for Applications in Nuclear Power Plants, IAEA-TECDOC-1511, IAEA, Vienna (2006).

[14]   INTERNATIONAL ATOMIC ENERGY AGENCY, A Framework for a Quality Assurance Programme for PSA, IAEA-TECDOC-1101, IAEA, Vienna (1999).

[15]   INTERNATIONAL ATOMIC ENERGY AGENCY, Attributes of Full Scope Level 1 Probabilistic Safety Assessment (PSA) for Applications in Nuclear Power Plants, IAEA-TECDOC-1804, IAEA, Vienna (2016).

[16]   INTERNATIONAL NUCLEAR SAFETY ADVISORY GROUP, Probabilistic Safety Assessment 75-INSAG-6, INSAG Series No. 6, IAEA, Vienna (1992).

[17] INTERNATIONAL ATOMIC ENERGY AGENCY, Approaches to Safety Evaluation of New and Existing Research Reactor Facilities in Relation to External Events, IAEA Safety Reports Series No. 94, IAEA, Vienna (2019).

[18] INTERNATIONAL ATOMIC ENERGY AGENCY, Safety Reassessment for Research Reactors in the Light of the Accident at the Fukushima Daiichi Nuclear Power Plant, IAEA Safety Reports Series No. 80, IAEA, Vienna (2014).

[19] INTERNATIONAL ATOMIC ENERGY AGENCY, Safety Assessment for Research Reactors and Preparation of the Safety Analysis Report, IAEA Safety Standards Series No. SSG-20 (Rev. 1), IAEA, Vienna (2022).

[20] INTERNATIONAL ATOMIC ENERGY AGENCY, Guide on Incident Reporting System for Research Reactors, IAEA, Vienna (2011).

[21] UNITED STATES NUCLEAR REGULATORY COMMISSION, PRA Procedures Guide, A Guide to the Performance of Probabilistic Risk Assessments for Nuclear Power Plants, Vols 1 and 2, NUREG/CR-2300, American Nuclear Society, LaGrange Park, IL (2016).

[22] UNITED STATES NUCLEAR REGULATORY COMMISSION, Probabilistic Safety Analysis Procedures Guide, NUREG/CR-2815, USNRC, Washington, DC/Brookhaven National Lab, NY (1985).

[23] UNITED STATES NUCLEAR REGULATORY COMMISSION, Handbook of Human Reliability Analysis with Emphasis on Nuclear Power Plant Applications, NUREG/CR-1278, USNRC, Washington, DC (1983).

[24] ELECTRIC POWER RESEARCH INSTITUTE, Systematic Human Action Reliability Procedure, EPRI-NP-3583, EPRI, Palo Alto, CA (1992).

[25] INTERNATIONAL ATOMIC ENERGY AGENCY, Case Study on the Use of PSA Methods: Human Reliability Analysis, IAEA-TECDOC-592, IAEA, Vienna (1991).

[26] INTERNATIONAL ATOMIC ENERGY AGENCY, Collection and Classification of Human Reliability Data for Use in Probabilistic Safety Assessments, IAEA-TECDOC-1048, IAEA, Vienna (1998).

[27] UNITED STATES NUCLEAR REGULATORY COMMISSION, Fault Tree Handbook, NUREG/CR-0492, USNRC, Washington, DC (1981).

[28] INTERNATIONAL ATOMIC ENERGY AGENCY, Use of a Graded Approach in the Application of the Safety Requirements for Research Reactors, IAEA Safety Standards Series No. SSG-22, IAEA, Vienna (2012).

[29] UNITED STATES NUCLEAR REGULATORY COMMISSION, Alternative Radiological Source Terms for Evaluating Design Basis Accidents at Nuclear Power Reactors, Regulatory Guide 1.183, USNRC, Washington, DC (2000).

[30] INTERNATIONAL ATOMIC ENERGY AGENCY, Case Study on Assessment of Radiological Environmental Impact from Potential Exposure, IAEA-TECDOC-1914, IAEA, Vienna (2020).

[31] INTERNATIONAL ATOMIC ENERGY AGENCY, Prospective Radiological Environmental Impact Assessment for Facilities and Activities, IAEA Safety Standards Series No. GSG-10, IAEA, Vienna (2018).

ANNEXES: SUPPLEMENTARY FILES

The annexes are available as on-line supplementary files and can be found on the individual web page of this publication at www.iaea.org/publications.

ABBREVIATIONS

CDF	core damage frequency
CET	confinement/containment event tree
FMECA	failure mode, effects and criticality analysis
FOS	facility operational state
LOCA	loss of coolant accident
LPSD	low power and shutdown
PIE	postulated initiating event
PSA	probabilistic safety assessment
RAM	reliability, availability and maintainability
SSCs	systems, structures and components

CONTRIBUTORS TO DRAFTING AND REVIEW

Amri, A.	Organisation for Economic Co-operation and Development, Nuclear Energy Agency
Barnea, Y.	International Atomic Energy Agency
Böck, H.	Atominstitut, Austria
Brinkman, H.	Nuclear Research and Consultancy Group, Netherlands
Corenwinder, F.	Institute for Radiological Protection and Nuclear Safety, France
D'Arcy, A.	South African Nuclear Energy Corporation, South Africa
Dybach, O.	International Atomic Energy Agency
Kowal, K.	National Centre for Nuclear Research, Poland
Leiva, C.	INVAP, Argentina
Oliveira, P.	Nuclear and Energy Research Institute, Brazil
Poghosyan, S.	International Atomic Energy Agency
Rao, D.V.H.	International Atomic Energy Agency
Sharma, R.	International Atomic Energy Agency
Shokr, A.M.	International Atomic Energy Agency
Sun, K.	International Atomic Energy Agency
Varde, P.V.	Bhabha Atomic Research Centre, India
Wijtsma, F.J.	Nuclear Research and Consultancy Group, Netherlands

Consultants Meetings

Vienna, Austria: 16–20 April 2012, 8–10 October 2012, 10–14 June 2013

IAEA
International Atomic Energy Agency

ORDERING LOCALLY

IAEA priced publications may be purchased from the sources listed below or from major local booksellers.

Orders for unpriced publications should be made directly to the IAEA. The contact details are given at the end of this list.

NORTH AMERICA

Bernan / Rowman & Littlefield
15250 NBN Way, Blue Ridge Summit, PA 17214, USA
Telephone: +1 800 462 6420 • Fax: +1 800 338 4550
Email: orders@rowman.com • Web site: www.rowman.com/bernan

REST OF WORLD

Please contact your preferred local supplier, or our lead distributor:

Eurospan Group
Gray's Inn House
127 Clerkenwell Road
London EC1R 5DB
United Kingdom

Trade orders and enquiries:
Telephone: +44 (0)176 760 4972 • Fax: +44 (0)176 760 1640
Email: eurospan@turpin-distribution.com

Individual orders:
www.eurospanbookstore.com/iaea

For further information:
Telephone: +44 (0)207 240 0856 • Fax: +44 (0)207 379 0609
Email: info@eurospangroup.com • Web site: www.eurospangroup.com

Orders for both priced and unpriced publications may be addressed directly to:
Marketing and Sales Unit
International Atomic Energy Agency
Vienna International Centre, PO Box 100, 1400 Vienna, Austria
Telephone: +43 1 2600 22529 or 22530 • Fax: +43 1 26007 22529
Email: sales.publications@iaea.org • Web site: www.iaea.org/publications